ELECTROPLATING

For the Metallurgist, Engineer and Chemist

by

J. B. Mohler

Director of Research
The Johnson Bronze Company

and

H. J. Sedusky

Research Electrochemist
The Cleveland Graphite Bronze Company

1951

CHEMICAL PUBLISHING CO., INC.

New York *N. Y.*

PRINTED IN THE UNITED STATES OF AMERICA

Contents

Introduction

The metallurgist and engineer are well acquainted with electrodeposition as a means of refining metal. This process, known as electrorefining, became widely accepted because of the high purity of the metal it produced. The very term "electrolytic copper" has become synonymous with metal of high purity. The metallurgist, in becoming familiar with this process, has learned that a certain metal will tend to go into solution, under the influence of current, exclusive of other metals in the anode. Thus, if an anode containing copper and gold is electrolyzed in a sulfuric acid electrolyte, the copper will go into solution and the gold will remain behind as an anode sludge. He has also learned that a certain metal will deposit, under the influence of current, exclusive of other metals in the electrolyte. Thus, if a copper sulfate electrolyte, such as would be obtained by the electrolysis previously mentioned, were to contain tin, the copper would still deposit exclusive of the tin. The copper deposit obtained by such a process would be rough and nodular but would be sufficiently sound to permit its use as a source of high-purity copper.

The process of electrorefining illustrates the fundamental requirements for electrodeposition; namely, an anode or positively charged electrode, a cathode or negatively charged electrode, an electrolyte or means of carrying current and a source of current. If these fundamental requirements are met, electrodeposition may take place, but the resulting deposit will not necessarily be of value.

If a metal salt is picked at random and current is applied to a solution of the salt, the results obtained will be varied. If the salt solution contains sodium chloride, only hydrogen will develop at the cathode. If a solution contains lead acetate, lead

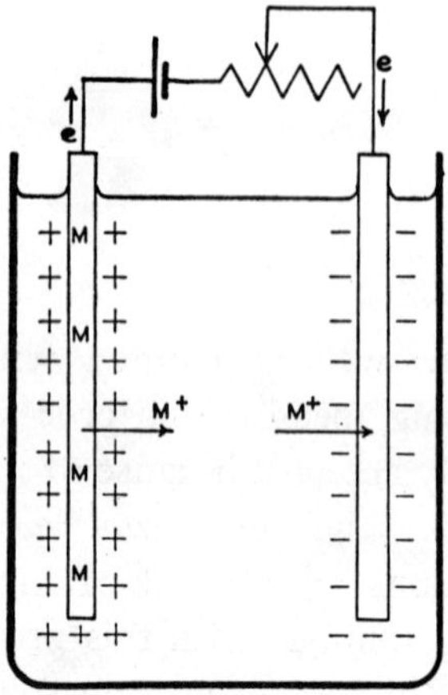

FIGURE 1
Electrodeposition

will deposit, but the deposit will appear as long crystals extending into the bath. If a solution contains stannic sulfate, the stannic ions will be reduced to stannous ions at the cathode and, at least for a short time and at a low current density, neither tin nor hydrogen will deposit. If a solution contains chromic acid, only hydrogen will develop; but if a small amount of sulfuric acid is added to the electrolyte, then chromium will deposit in addition to hydrogen.

By experimentation with solutions of metal salts, baths can be developed that will produce satisfactory deposits. A great many experiments are usually required and often it is necessary to study the effect of a host of organic substances known as addition agents. By experimentation and study, new baths have been developed to fill specific needs. For example, a deposit that is satisfactory for electrorefining may not be satisfactory for electroforming. A deposit that is satisfactory for electroforming may not be satisfactory for electroplating.

Electroplating is a process of electrodeposition by which a thin, smooth, sound metallic deposit is produced over a basis metal. This definition sets electroplating apart from the other processes of electrodeposition even though the requirements of the definition are not met in every case.

In this book, the fundamentals of electroplating will be briefly considered. A number of plating baths will then be discussed. The book is written for the trained metallurgist, engineer and chemist, with a good background in science but only a slight acquaintance with electroplating. Thus, we intend to go into more detail on the fundamentals than is required for an experienced plater or an electrochemist but considerably less detail than is required for one trained without a general knowledge of science.

Chapter 1

The Mechanism of Electrodeposition

The Electrolyte

The process of electrodeposition is one in which electric current is carried across an electrolyte and in which a substance is deposited at one of the electrodes.

The electrolyte is the medium that carries the current by means of ions. The ability of a solvent, especially water, to ionize substances dissolved in it, *i.e.,* to split them into components, that carry positive and negative charges makes electrolysis possible. The electricity is carried across the electrolyte by the charged ions, and products of electrolysis appear at the electrodes. This is a result of the positively charged ions being attracted to the negatively charged cathode while the negatively charged ions travel toward the positively charged anode. The charges of the ions are then neutralized by the charges on the electrodes and the products of the electrolysis appear at the electrodes.

The electrolyte is a conducting medium in which the flow of electric current is accomplished by the movement of matter. It is usually a substance that gives rise to ions on dissolution in a solvent, but a fused salt may also act as an electrolyte. If more than one ion is present, carrying a positive charge, several reactions are possible at the negatively charged cathode, although usually only one product of electrolysis appears. Each electrode reaction takes place at a specific voltage and the most positive metal ion will deposit at the cathode.

Any liquid or solution that contains ions can be used as an electrolyte. The large majority of commercial electrolytes, however, use water as the solvent and are, therefore, called aqueous electrolytes. Fused salts, which are a class of nonaqueous electrolytes, find their greatest use in the electrolytic production of metals such as sodium, magnesium and aluminum. Fused-salt electrolytes are also used in the electrolytic cleaning of metals. Another type of nonaqueous electrolytes are solutions in organic solvents. These have been used for experimental electroplating of aluminum but have not as yet been applied commercially.

The extensive use of water as the solvent in the electroplating industry is due to its cheapness and abundance and to the fact that practically all the common salts used in the process are very soluble in water.

In this book, the aqueous electrolytes alone will be considered. The term electroplating will be occasionally simplified to plating, which is known to cover electroplating.

The Cathode

The plater is primarily interested in the reaction that takes place at the cathode since this is where deposits are produced. The potential at which this reaction takes place is called the deposition potential. This potential can be measured readily in the laboratory, but it is neither convenient nor desirable to measure it in the plating tank. The reaction at the cathode is much easier to follow by a knowledge of the quantity of current that reaches the cathode. If, in addition, the plater has some knowledge as to the distribution of the current over the cathode, he may then have some idea of how the plated article will be coated with metal. He will be able to predict the time required to produce a desired thickness of deposit and also have an idea of how the thickness will vary from one area to another of the plated piece.

Unfortunately, it is difficult to make accurate predictions regarding the distribution of metal over the cathode. In practice, the quantity of current is controlled and the current is allowed to flow for a definite period of time after which the local thickness of metal on the plated piece is measured. If an undesirable distribution of metal is obtained, adjustments are made in the racking or positioning of the pieces or of the anodes. In some cases, chemical adjustment of the electrolyte may improve the metal distribution.

The current allowed to flow to the cathode is in proportion to the area being plated, so that the current is expressed as current density or quantity of current per unit area. In commercial plating, the current density is expressed as amperes per square foot.

Occasionally, a plating tank is controlled by voltage. This procedure is less satisfactory than control by current density, since the tank voltage is affected by many factors other than the reaction taking place at the cathode. Tank voltage, however, is very easily measured and often gives information about changes in the plating process, such as a reduction in conducting salt content, or polarization of the anode. Plating control by tank voltage is satisfactorily used in cases where the cathode area is difficult to measure, such as in barrel plating.

The Anode

The reactions taking place at the anode are almost independent of the reactions occurring at the cathode. The position of the anodes naturally has much to do with the distribution of current at the cathode, but the anodes usually operate best at a range of current densities that can be changed independently of the cathode current density by changing the anode area.

The Balanced Bath

A plating bath can be operated successfully for long periods of time if the composition of the bath is not changed too rapidly. Such changes are primarily due to:

1. Chemical decomposition
2. Incomplete electrode reactions
3. Drag-in or drag-out

The bath stability can best be illustrated by consideration of several typical baths.

An acid copper bath is relatively easy to control because there is very little tendency for chemical decomposition and the reactions at the electrodes are essentially complete. This means that the electrolyte is chemically stable and that for every chemical equivalent of copper dissolved at the anode, there is a chemical equivalent of copper deposited at the cathode. Nevertheless, the bath cannot be continuously and satisfactorily operated without control, since solution is lost from the system by drag-out every time a rack is removed. Control is further complicated in that glue is generally added to the bath to produce a finely crystalline deposit. The glue is not stable and must be closely controlled. This bath is typical of many acid baths where the major factor in control is the addition agent used; in this case, the glue.

An alkaline tin bath is not a typical alkaline bath, but it is a good example of a bath where care is required to maintain solution balance.

In the alkaline tin bath, the electrode reactions are not complete, i.e., the anode and cathode efficiencies are less than 100 per cent. In addition to depositing tin, hydrogen is evolved at the cathode. And in addition to tin being dissolved, oxygen is given off at the anode. To make the problem more complex, the bath undergoes continuous chemical decomposition and the anodes must be maintained with an oxide film at all times. If the oxide film is not present, or is too thick, troubles set in that throw the bath out of balance or even cause a deposit to form that is not acceptable. The bath can be controlled by regulation of bath temperature, cathode-current density, anode-current density and by chemical analysis and proper chemical additions. The bath voltage is responsive to changes at the anodes; and since the bath is sensitive to anode changes, the voltage may be used as an aid to bath control.

A plating bath should always be kept within prescribed chemical limits whether or not the bath composition is difficult to maintain. The bath should also be used in such a way that a minimum of chemical additions is required. It is very rare that a bath does not require additions for long periods, although this condition is approached where the anode efficiency is slightly higher than the cathode efficiency. Such a bath would be a perfectly balanced bath, but even in this case, drag-out would eventually remove a sufficient amount of one of the essential chemicals so that chemical additions would be required. Since additions are required to all baths, it is best to make them frequently and in small amounts, so that chemical limits are easily held. Addition of large quantities of chemicals to the plating bath often leads to trouble. For example, for a large addition of chemicals, a small error in chemical analysis is liable to result in a concentration exceeding the chemical limits on the bath. Moreover, the chemicals often contain impurities that are not harmful for small additions, but that require electrolysis with dummy cathodes before the bath may be used, if the additions are large.

The Equilibrium Potential

It is well to settle a few points regarding the equilibrium potential before considering some of the details of the plating process.

The well-known series of potentials for electrochemical reactions is shown in Table 1. This series has also been called the electrochemical series or the electromotive force series of elements. It is a reliable point for reference; however, it can be highly misleading if it is used as a definite guide. These potentials are equilibrium potentials. They were measured under conditions where no current was flowing. They were measured for definite quantities of dissolved salt present and at a standard temperature to obtain comparable valves.

Hydrogen is taken as the reference point on the electrochemical scale and is arbitrarily assigned a value of zero. The metal ions whose voltage is listed as positive are more reactive than the hydrogen ion when they are present in equivalent chemical quantities, whereas the metal ions having a negative voltage are less reactive than the hydrogen ion. Thus as we go up the scale, the metal ions become more and more reactive at the cathode, i.e., they become more electropositive (attracted to the cathode), or they deposit more readily. As we go down the scale to more negative electrode potentials, the metals become

TABLE 1

Electrochemical Potential Series

Ion	*Voltage*
Au^+	+1.5
Ag^+	+0.7995
Cu^+	+.0528
Sn^{++++}	+0.003
H^+	0
Pb^{++}	−0.1264
Sn^{++}	−0.1406
Ni^{++}	−0.231
In^+	−0.336
Cd^{++}	−0.4024
Cr^{+++}	−0.509
Zn^{++}	−0.762
Na^+	−2.7125

more electronegative, or they go into solution more readily (anodic metals).

With reference to this scale, it has been said that any metal on the scale will displace from solution all those metals that appear above it. This is true for the exact conditions under which the potentials were measured, but it cannot be taken as a general rule. According to this rule, if a piece of copper is immersed in a solution containing silver ions, silver will deposit on the copper. This will occur in acid solutions even though the solution contains a very small quantity of silver. However, the opposite reaction can be made to take place. That is, if silver is immersed in a concentrated acidic solution of copper, copper will deposit on the silver. In other acid solutions tin can be deposited on copper [1] even though tin is above copper on the electrochemical scale. These apparent exceptions occur because the conditions under which the experiments are carried out differ greatly from those under which the potentials were measured. However, the metals in the series are arranged according to their relative reactivity. The metals below hydrogen can be dissolved readily in acids and those very low in the table will react with water. The metals above hydrogen can only be dissolved in acids under oxidizing conditions. Thus the table can be used for common cases, but if it is to be applied to new or unusual conditions, it will be necessary to experiment or measure the potential under the new conditions. If the potential is measured during plating, it is measured under dynamic conditions and is called a deposition potential.

The Deposition Potential

The potential of an electrode and of a solution of its ions may be measured during plating. This deposition potential varies with the concentration of metal ions in the bath, and is also greatly affected by the current density. As the current density is increased, polarization at the electrode increases, resulting in conditions more favorable for deposition of metals low in the electrochemical scale. Thus, it becomes possible to deposit zinc from acid solutions, whereas zinc normally dissolves in acids. One might expect that zinc could not be deposited in the presence of hydrogen ions, and under some conditions it is difficult to achieve this. If polarization does not exceed the hydrogen overvoltage, hydrogen will develop exclusive of zinc. This can take place even during the electrolysis of copper from a copper nitrate solution. Such an electrolysis is carried out in the presence of nitric acid, and by continued electrolysis, the copper can be deposited completely from the solution. However, if the solution is heated while copper is depositing, a point will be reached where the tendency for the copper to dissolve will be greater than its tendency to plate. At this point, the copper will go back into solution in the nitric acid even though the current is flowing. The behavior of copper in nitric acid solutions and that of zinc in acid solutions illustrate that, for a study of the mechanism of deposition, the measurement of deposition potentials is necessary. These potentials direct attention to the importance of polarization and overvoltage.

Reference

1. U. S. Patent 2,369,620.

Chapter 2

The Laws and Characteristics of Plating Baths

The plater is concerned with the quantity of electricity that is used in his process, since the number of units of electricity that pass through the electrolyte during electrolysis determines the quantity of product formed at the cathode.

The unit quantity of electricity is the coulomb which is the quantity of current passing in a circuit when 1 ampere flows for 1 second. Ordinarily, the ampere rather than the coulomb is considered in electroplating. The ampere, however, gives only the current strength and requires a time factor so that the quantity of metal obtained at the cathode is calculated from the ampere-seconds or ampere-hours.

One gram equivalent weight of ions will be discharged at the cathode by 1 faraday. One faraday equals 96,500 coulombs, 96,500 ampere-seconds, 1608.4 ampere-minutes or 26.806 ampere-hours.

One gram equivalent weight of a metal equals one gram atomic weight divided by the valence. The valence is equal to the number of charges on the ion. From the faraday, the grams deposited at the cathode can be calculated as in Table 2.

The quantity of electricity that passes through the electrolyte during electrolysis is important because a minimum amount of metal is required for every application; and also it may not be desired to deposit an excess of metal for economic or specification purposes. However, it is not enough to know the total quantity of current that passes. The distribution of the current and the plating rate must also be known for a reasonable estimate of the plate thickness. If the piece to be plated is a large flat sheet, then 10 to 20 per cent of the current

TABLE 2

Calculation of the Grams of Metal Deposited at the Cathode

Metal	*Ion*	*Gram Atomic Weight*	*Valence*	*Grams Deposited per Faraday*
Silver	Ag^{+}	107.88	1	107.88
Copper	Cu^{+}	63.57	1	63.57
Copper	Cu^{++}	63.57	2	31.78
Gold	Au^{+}	197.20	1	197.20
Tin	Sn^{++++}	118.70	4	29.67

may be lost to the edges. The current density over most of the area will then be 10 to 20 per cent lower than the current density obtained by dividing the total current by the total area. From this approximate current density, the approximate plating rate can be calculated. In many cases, however, the cathode efficiency is not 100 per cent so that some percentage of the total current is consumed in evolving hydrogen.

The cathode efficiency is that percentage of the current that deposits metal. In other words, if 60 per cent of the current deposits metal, then the plating rate is only 60 per cent of that which would be obtained by calculation from the ampere-hours. Just as it is more convenient to use current density than total current, it is also convenient to use ampere-hours per square foot to calculate plating thickness. Of course, if the cathode efficiency is less than 100 per cent, then the plating rate is calculated from the following equation:

$$\text{plating rate} = \frac{\text{amp.-hr.}}{\text{sq. ft.}} \times \%\ \text{efficiency} \qquad (1)$$

The plating rate calculated from equation (1) is in ampere-hours per square foot. In order to convert this to thickness, the chemical equivalent and the specific gravity of the metal must be taken into consideration. This can be done with the following formulas:

$$\frac{\text{equivalent weight}}{\text{amp.-hr. faraday}} = \frac{\text{equivalent weight}}{26.8} = \frac{\text{g.}}{\text{amp.-hr.}} \text{ at } 100\% \text{ efficiency} \tag{2}$$

$$\frac{\text{g.}}{\text{amp.-hr.}} \times \frac{\text{amp.-hr.}}{\text{sq. ft.}} = \frac{\text{g.}}{\text{sq. ft.}}$$

$$\frac{\text{g.}}{\text{sq. ft.} \times 144} \times \frac{1{,}000}{16.4 \times \text{sp. gr.}} = \text{mils thickness} \tag{3}$$

These formulas illustrate how the total current produces a total thickness. In every application, it is convenient to know the ampere-hours per square foot and the cathode efficiency. Not only is it convenient to have this information for the average areas, but it is also desirable to have it for the areas of highest and lowest current density. If the distribution of current is known and if a specification for thickness is given, then the quantity of current required can be determined.

Unfortunately, the distribution of a deposit over the surface of an article is not easily estimated and in many processes, it is impossible to calculate. Nor can the current density always be calculated and the cathode efficiency often changes too rapidly to make cathode-efficiency measurements worth while. This is especially true of barrel-plating operations or in applications in which the amount of current being taken by the plating rack is unknown. The plating rate then *cannot* be calculated but must determined by measurement of the plate thickness, after plating for a certain period of time at a total current per rack or per barrel.

Experimentation in the plating room to obtain proper metal distribution should be guided by a knowledge of the qualitative rules for the behavior of the cathode.

These qualitative rules may be stated as follows:

1. The current concentrates on corners and edges.
2. The current does not penetrate into deep recesses.
3. The current does not readily flow around a nonconductor.
4. One conductor will rob another of current.

These rules may be used as a guide for the arrangement or geometry of a plating system.

Electrode Position

Rule 1 tells us that excessive current may be lost to projections or edges of the article. We also know that current will tend to follow the shortest path between conductors. This fact can be taken advantage of in order to reduce current concentration on edges by placing the edges farthest from the anode. Thus in plating from an anode to a sheet, the system of Figure 2 can be rearranged to that of Figure 3.

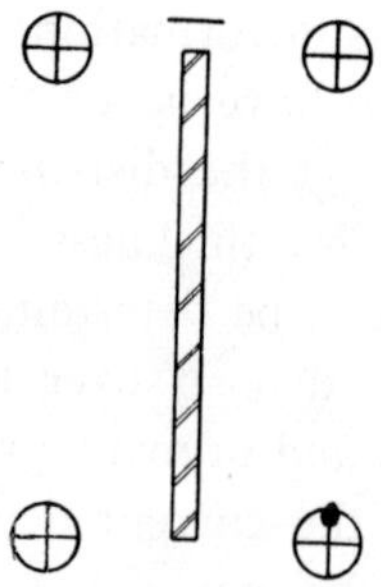

FIGURE 2
Anode Arrangement to Favor Plating on Corners

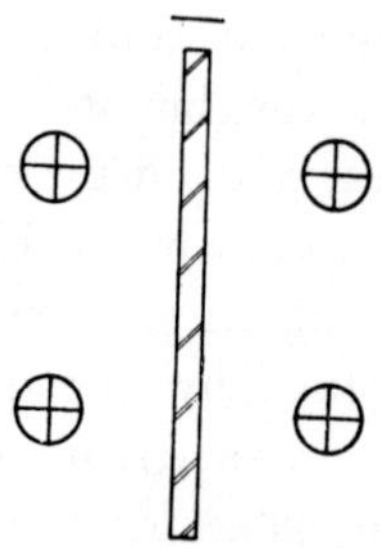

FIGURE 3
Anode Arrangement to Minimize Plating on Corners

According to rule 2, if it is desired to plate inside of a large recess, the obvious thing to do is to place an anode in the recess. The anode should be shaped so that the anode-to-cathode distance is the same at all points over the surface and a little longer to edges.

Shadowing

A problem cannot always be solved by the proper arrangement of the system. Even when a solution is possible, it is not always economical. Anodes are difficult to support in fixed position with respect to the cathode and they change in size as they are used.

The edges of a sheet may be shadowed by the use of a nonconductor. The shadow does not have to be between the anode and the cathode but merely protect the cathode from receiving current in more than one direction. If a flat sheet cathode is placed between two nonconducting walls, the distribution of current over the cathode will be almost perfect, as shown in Figure 4.

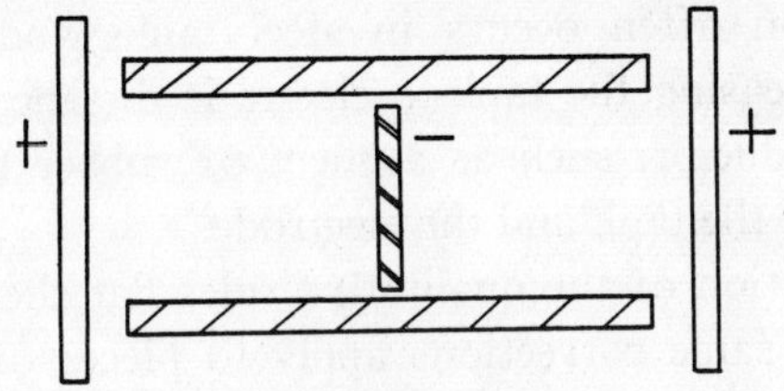

FIGURE 4
Shadowing

The walls prevent the current from coming from any other direction than in straight lines within the boundaries of the insulating walls.

Robbing

If several flat sheets are plated at the same time, one may be used to rob the edges of the next. Fair uniformity may be obtained by an arrangement as shown in Figure 5.

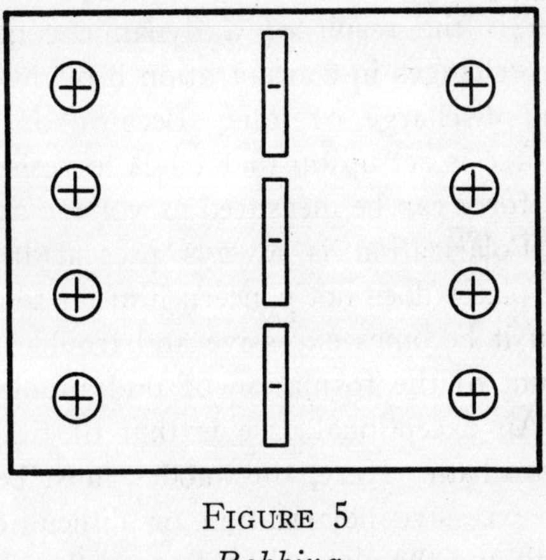

FIGURE 5
Robbing

Here the sheets are close together and the edge of one robs current from the next. The edges next to the tank are shadowed by the walls of the tank provided the tank is a nonconductor. If the tank is a conductor, then the shortest electrical path may be from the anode to the tank and from the tank to the cathode. This condition often occurs in steel tanks and is corrected either by increasing the tank-to-electrode distance or by hanging a nonconductor, such as a sheet of rubber or a sheet of glass, between the tank and the electrode.

For illustration of the qualitative rules flat sheets have been used, but the same corrections apply to pieces of complicated design where they perform just as well. High-current-density areas should be robbed or shadowed. Corrections should be made to get more current to low-current-density areas either by making the anode-to-cathode distance more favorable or by shadowing the high-current-density areas to such an extent that a greater percentage of the current reaches the low-current-density areas.

Polarization

Polarization is a term that is widely used by the electroplater, for there are many electrode reactions that can be explained through polarization.

Polarization is the result of a dynamic condition at the cathode, due to changes in concentration brought about by the movement and discharge of ions. Because of the dynamic condition, a force is set up up that tends to resist the flow of current. This force can be measured as voltage and it is called polarization. Polarization is always present in electrolysis. Normally, the plater does not concern himself too much about it except when it becomes excessive and troubles are encountered on account of the formation of undesirable products at the cathode. An exceptional case is that of tin anodes in a sodium stannate bath. Here, the anodes must be maintained in a state of excessive polarization or difficulties will arise in both the bath and the plate. In this case, the term excessive polarization is sometimes used, although it is more common to say polarization when one really means excessive polarization.

The electrode reactions can be best explained by the use of an ideal current-voltage curve, such as shown in Figure 6.

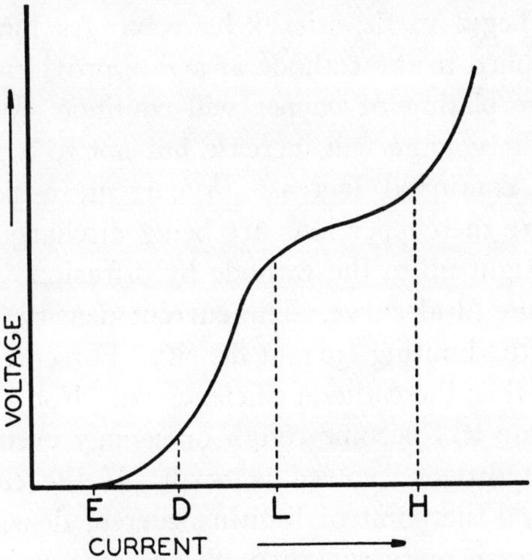

FIGURE 6

Current-Voltage Curve

If the voltage from the cathode to appoint at a very short distance from the cathode is measured under ideal conditions, a curve as shown in Figure 6 is obtained. At point E, no current is flowing so that the voltage is that of the equilibrium potential. Now, if a small amount of current is applied to the system, the voltage will increase, or, in other words, polarization will begin to take place. However, the deposition of metal will not yet occur even though there are platable ions in the aqueous solution. As the current is further increased, the voltage also increases until finally a voltage is reached where plating takes place. This voltage is point D which is known as the decomposition potential or the point at which the electrode reactions begins. If the solution consists of water and sulfuric acid, bubbles of hydrogen gas will appear at the cathode at this point and the voltage, in this case, is called the hydrogen overvoltage. Hydrogen overvoltage is merely a form of polarization.

Now if the solution contains sulfuric acid and copper sulfate in water, a decomposition potential will be reached where copper will begin to deposit exclusively. As long as copper ions are supplied to the cathode at a rate providing an excess, this exclusive plating of copper will continue. With increase in current, the voltage will increase but not to a great extent. But with a continued increase in current, a point will be reached where the copper ions are being discharged as fast as they are brought up to the cathode by diffusion. This will be point L on the ideal curve. The current density at this point is known as the limiting current density. For current densities greater than this, the cathode efficiency will drop and, in many cases, the plate will become rough or spongy even though the amount of hydrogen evolved is small. If the current is increased beyond the point of limiting current density, the voltage will increase more rapidly with increase in current until the point H is reached. At point H, hydrogen will be evolved rapidly and we have now exceeded the hydrogen overvoltage with the copper ion still present but not in sufficient concentration to deposit exclusively. Beyond the point H, the voltage will not increase greatly with increase of current. This is due to the aqueous solution having a very great capacity to supply hydrogen ions as fast as they are discharged.

Plating Quality

The quality of plating varies with the application. If the only purpose of the deposit is to prevent rusting of steel for a short period of time, then a very thin coating of zinc may be sufficient. In this application, the deposit will be a quality deposit if it completely covers the piece. If it is further required that the deposit is bright, then the job of quality plating becomes a little more exacting.

The quality of thin deposits may be controlled by testing

for thickness and corrosion resistance. As long as representative samples pass these tests, the specifications are met. The specifications, however, become more rigid for more exacting applications. For example, a heavy deposit for an aircraft bearing must be almost flawless. The deposit must be bonded to the steel back and it must be practically free from pores down to a microscopic size. The finished bearing must pass strict tests for porosity and adherence. Here quality has a meaning entirely different from that for flash deposits. And to produce the desired quality, very close control is required throughout the entire plating cycle.

In an important engineering application, quality may only be obtained by operating a plating bath well above the decomposition potential and well below the limiting current density. It may further be required that the bath is continuously agitated and heated, and that temperature, chemical limits and the geometry of the plating system are closely controlled. Wide limits may be possible in some of the bath variables, but only if one variable is changed at a time. Therefore, close limits are set on all variables so that if all change in an unfavorable direction at the same time, the bath will still produce quality plating. In order to set limits for a strict plating problem, some knowledge of plating range is required.

Plating Range

The plating range is the range of current density over which the deposit will be satisfactory. It is usually a range of current densities somewhat above that at the decomposition potential and somewhat below the limiting current density. It will be very wide in an alkaline tin bath and quite narrow in a chromic acid bath.

In Figure 7, plating takes place from a bar anode to a sheet where the edge of the sheet faces the anode. A section of the system is shown for a deposit from an acid bath of fair throwing power. On the edge of the sheet facing the anode, the deposit is rough. The current density on this edge is very

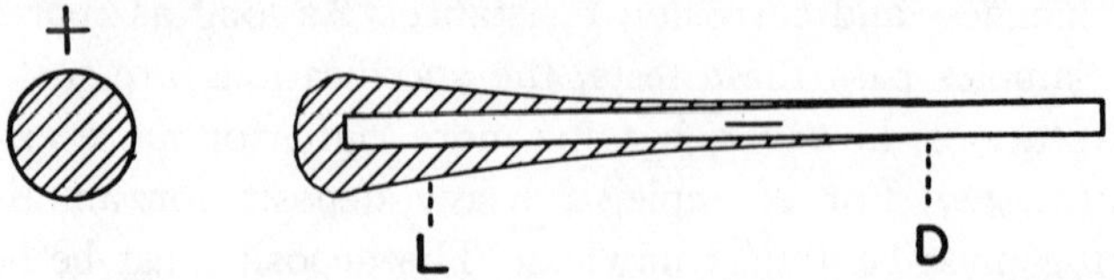

FIGURE 7
Plating Range

high and becomes less at greater distances from the anode. Point L is the limiting current density and point D is the current density at the decomposition potential. The deposit is often rough, crystalline, off color or spotty at current densities that are very high or very low. Below some maximum and above some minimum current densities, the deposit is satisfactory and the range between these densities is the plating range.

The plating range can be measured experimentally by calibrating an arrangement of the anode and cathode, and using a number of such arrangements for plating-test control. The two most common of these tests are the bent-cathode test [1] and the Hull cell test.[2]

REFERENCES

1. Pinner and Baker, *Trans. Electrochem. Soc.*, 77, 353 (1929).
2. F. MacIntyre and R. O. Hull, *Convention Proc. Am. Electroplaters Soc.* (1943).

CHAPTER 3

The Control of a Plating Bath

A plating bath can be controlled by holding all chemical ingredients within specified limits. It can also be controlled by plating-range tests and, in a few cases, it can be controlled with a hydrometer and a simple test.

The control method selected depends primarily on the type of deposit desired. Baths that are used for flash deposits usually may be run for long periods without the use of tests or analyses. For instance, an alkaline tin bath may be operated for months at a time by merely using steel anodes and making frequent additions of sodium stannate based on the total current applied to the tank. However, if a high-efficiency copper cyanide bath is used for a deposit of 1 mil of copper, the bath must be carefully controlled by chemical analyses biweekly.

For each application then it is necessary to decide on control methods. The most frequently used and most reliable methods are those of analytical chemistry. If chemical limits are set on a bath and are proved to be adequate for a specific application, then the question of the bath control is merely a matter of a few analyses. If the bath is analyzed and adjusted within chemical limits, attention can be paid to other plating variables such as temperature, racking and the preparatory steps taken prior to plating. It is convenient to the plater to know that his bath is in the proper limits.

Good analytical methods are available for most of the baths, but in many cases the choice of methods is rather limited. Methods that are rapid and of sufficient accuracy are not available to determine the impurities that enter the bath: and the impurities must be held below specified limits just as the chemical ingredients must be held within limits. If plating is being carried out on zinc-base die castings, zinc will build up in the bath since it is an active metal. When zinc is present in excessive quantity, it will cause trouble in most of the baths. Metal impurities regularly enter the baths by their reaction with the plating racks and the metal to be plated. In addition, metal may be dragged in from the preparatory tanks.

Small amounts of lead will ruin the deposit from almost all of the alkaline or cyanide baths. Lead, tin, zinc and chromium will cause trouble in a cyanide copper bath. Copper, zinc, chromium and iron will contaminate a nickel bath, and any of the noble metals may cause trouble in an acid lead or acid tin bath. A complete list of metals that cause trouble would be difficult to compile because the sensitivity of the bath to the impurity changes with change in bath composition. The metals to test for are those that are present either as metals or as metal salts in the plating room.

There are no adequate analytical methods for the determination of metal impurities, therefore, it is obvious that chemical control cannot be completely relied upon. Also it will often be found that a bath will not operate properly even though it is within the given chemical limits. This merely means that something is causing trouble that is not included in the analyses. The cause may be something that cannot be determined by the usual analytical methods.

The cause may be the presence of an organic compound whose detection by analytical methods may be extremely difficult or no method may be available for its determination. Even the known organic substances that are added to the bath often cannot be controlled by analysis.

Although the analytical methods are most reliable for general plating-bath control, they do not guarantee proper plating. Therefore, a plating test is needed. The simplest plating test consists of plating on a flat sheet of metal or on one of the articles regularly plated in the bath. However, such tests only detect trouble after it has occured; and the primary purpose of testing and analysis is to keep out of trouble.

If a plating-range test is employed, then trouble may be forestalled. This is true because troubles often start at either high or low current densities and creep into the normal plating range. The plating-range test, such as the Hull test or the bent-cathode test, pick up troubles at the current densities outside the normal plating range and corrective steps can be taken before trouble occurs at the plating tank.

There are very few baths that cannot be controlled by chemical analysis and a plating-range test. However, these tests are not required in every case and they are not necessarily adequate in every case. Let us assume that we have a fully automatic plating machine where the time and the current in the plating tank is limited and the thickness of the deposit is specified. In this case, it will also be necessary to control the plating rate. Chemical analysis, temperature control and current control do not guarantee the cathode efficiency. The cathode efficiency may drop as impurities enter the bath even though these impurities do not harm the quality of the deposit or are not revealed in a plating-range test.

The plating rate is measured either in terms of thickness of metal deposited in a given time or in terms of cathode efficiency at a given current density. If it is measured as thickness, it may be measured on a test panel; but more often, it is measured directly on one of the plated articles. Such measurements may be made by magnetic means with a proper combination of metals, or chemically. The usual chemical methods consist of determination of the time necessary to dissolve the deposit, or stripping the deposit from a known area, followed by analysis for the total metal stripped. Cathode efficiency is measured by plating a weighed test piece for a known time and current and reweighing to determine the amount of metal deposited.

The selection of analytical methods depends on the speed and accuracy required as well as where the method is to be run.

It is often convenient and sometimes necessary to carry out tests and analyses near the plating line. Also, it is often true that some of the ingredients can only be determined in a properly equipped analytical laboratory.

Most analytical methods that have been adopted to plating control are relatively simple, as analytical methods go, and can be run with a minimum amount of chemical equipment. A few, however, such as the determination of Rochelle salt, are relatively complex. These more complex methods must be run with proper equipment and by a trained analyst. Many of the methods, however, can be run at or near the plating line. Then too, it will be found that the simple methods are most successful when they are run in the plant. For rapid control methods, it is best to sacrifice some accuracy for the sake of simplicity. Furthermore, the closer the control method is to the line and the sooner the results are used for bath correction, the less need there is for a high degree of accuracy. The analyst who does not know what the tank is doing can only make required analyses and guarantee his accuracy from his knowledge of analytical chemistry. The analyst may not know how his results are used to correct the bath. The plater, on the other hand, may run a plating-range test and suspect low metal content. If he can confirm his suspicion by some approximate rapid check on the metal content, he can add metal and run a second plating-range test. This may be done in less time than is required for sampling and transferring the sample and the results to and from the laboratory. The plater, with a few simple tests, can often do a good control job; and it takes the mystery out of control work. If the plater has never seen a laboratory or does not understand the analytical procedures, he will not have sufficient confidence in the numbers he receives from the laboratory. Improper sampling can cause no end of trouble between the plant and the laboratory for the analyst can only analyze what he receives. In some cases, the plater may take a sample before his salts are completely in solution on making up a fresh bath. Many plating salts, such as boric acid, dissolve very slowly, and others, such as the alkalis, will form large slowly dissolving lumps at the bottom of the bath if they are not added properly. If the plater has a rapid control method at the line, he will learn what is happening when he makes up a new bath and when he makes additions. A rapid over-all check can be made on any fresh bath with a hydrometer. The fresh bath must contain some total salt concentration made up from specified quantities of various chemicals which will always be of the same specific gravity. The hydrometer is a valuable tool that should be used wherever possible because it is extremely rapid and, in proper applications, is highly accurate. After a bath has aged and accumulated impurities or chemical decomposition products, the hydrometer may be useless, but it will always apply to a fresh bath and in many cases, it gives reliable information throughout the life of the bath.

It is possible, for example, to control a chromic acid bath with nothing more than a hydrometer and a plating-range test.

The hydrometer is probably the simplest instrument that can be used for solution testing. There are other simple methods that can be used such as the measurement of the pH with pH papers and the measurement of the conductivity.

A conductivity measurement is similar to a gravity measurement in that it is rapid and is affected by all the salts present, but it does not measure the gravity. A conductivity meter responds principally to the hydroxyl ion in alkaline solutions and principally to the hydrogen ion in acid solutions, for these are the ions that carry most of the current. It is applicable to direct control of electrolytic alkali cleaners because current-carrying capacity is of prime importance in such a cleaner and the hydroxyl ion usually does a great deal of the cleaning work.

The conductivity meter may be used in conjunction with some other test. If an acid and a salt are present, then both can be determined by measurement of conductivity and gravity. The acid primarily determines the conductivity, but a correction must be made for the salt present. The gravity measures the total solids present. Now by setting up proper original standards, both the acid and the salt may be measured. This procedure works well where one substance is of high conductivity and the other of low conductivity. A temperature measurement must also be made, or all tests must be carried out at the same temperature, for the temperature has a considerable effect on both conductivity and gravity. Determination of the amount of acid and the amount of iron in a pickle solution is a typical example of this type of control.

It is possible to develop other rapid control methods, such as the measurement of cathode efficiency. In many of the alkaline baths, the cathode efficiency is low at high current density and the anode efficiency is high at low current density. Advantage may be taken of this by plating a cathode for a definite period of time at a high or at operating current density and then measuring the time for deplating the cathode at low current density. From the data obtained, the cathode efficiency at the operating current density can be readily calculated. In some baths, such as a silver strike, the cathode efficiency is a direct measure of the metal content and its measurement is thus a rapid method of determining the silver content of strike baths.

When unusual methods are used for solution control, the property that is being measured should be kept in mind. Thus conductivity may measure only the alkali content of a cleaner solution, but in an electrolytic cleaner, the conductivity may be a more important control factor than any of the chemical substances the analyst may determine. If cathode efficiency is used as a means of checking metal content, it may be more important than an analysis since the plating rate is being measured and a direct control of plating rate may be more important than that of the chemical content.

One can go far with simple nonchemical methods, but the simplicity of many of the chemical methods should not be disregarded. As an example, direct determination of free cyanide in a silver bath is a simple, rapid and accurate chemical method. It is doubtful if any other method is better than direct titration of free cyanide in a silver bath. Many other titrations, such as for free acid, are just as simple and quite easily run at the plating line in the same manner as in the laboratory. A few solutions and simple glassware are all that is required. However, many methods can only be run in a properly equipped laboratory.

The application of the bath has much to do with the selection of the best methods. A bath that is used for flash deposits may be controlled by simple methods, but a bath that is used for heavy, sound deposits may require strict, careful chemical analyses and physical tests.

Solution control is a must for successful plating. There is no plating bath that will adjust itself to the proper equilibrium by electrolysis. But solution control is not the whole answer. If too frequent additions are required for continuous quality plating, then the bath is too far from balance. A perfect balance cannot be maintained, but it should be approached. If the bath changes rapidly, an adjustment of anode and cathode areas or of current should be made until balance is approached. Great changes in bath composition should be avoided since great changes mean extensive adjustments and such adjustments involve sudden change in bath properties and a shift in the plating range. A large addition may require some other secondary correction and result in loss of production time. It is often helpful to keep a daily log of a bath, showing chemical composition, tests, additions, current density, production time, temperature and appearance of the bath and of the anodes. It is also revealing to plot these data against time, and it will often be found that the bath will change over a period of weeks and that at times this can be coordinated with the number of anodes or some other data. Changes from day to day are difficult to see, but a gradual change over a period of months is apparent from a plot of the log data. Such a plot may reveal that additions become much larger and corrections more frequent when the total current is increased. This may be due to the fact that the tank is not large enough for excessive plating. If a tank is crowded with work and run continuously, it will be more difficult to control than the same amount of work in a larger tank.

Overworking a tank may lead to rapid build-up of impurities. Impurities can enter the bath from everything that comes in contact with it. The racks, chemicals, the water, the air and the articles to be plated. The impurities are of two general types. Those that will plate and those that will not plate. The metals that plate more readily than the metal in the bath usually cause trouble. They cause the most trouble in low-current-density areas and that is where they will be revealed on a plating-range test. The plating-range test also gives the clew as to means of removal of the impurities. By electrolyzing the tank at low current density, troublesome noble metals are removed. Metals or other impurities that do not plate usually are not harmful unless they build up in large amounts. It will then be necessary to remove them by crystallization or by chemical means.

Plating-bath troubles are listed in chapter 24 and analytical methods are given in chapter 25. Control of a plating bath is usually accomplished by standard analytical methods, however, the control procedures for a certain application cannot be listed as they are best determined by experience and experimenting.

CHAPTER 4

The Deposit

An experienced plater can control a bath for long periods of time merely by observing the appearance of the deposit. He is quick to recognize the need for additions by appearance of dull areas on the plated work. He knows that red copper deposits mean low free cyanide in a copper bath. Furthermore, he knows the proper addition and current changes to bring about an immediate correction. Although such practice is good in emergencies, it is not recommended. But it does demonstrate that the product is the deposit and that it is possible, if not desirable, to continue to produce a good product independently of changes in the bath.

The Environment

The deposit is mostly affected by the environment in the immediate vicinity of the cathode. It is not affected by dirt that lies on the bottom of the tank or by sludge that adheres to the anode. It is affected by tank walls that are close to the work and also by suspended charged particles if they are close enough. With time, the plate may be affected by dirt or sludge that becomes suspended in the bath and reaches the deposit, but in a given tank, the condition at the cathode is somewhat independent of the composition of the bath and of the anodes. Identical deposits can be produced in baths that are entirely different by changing the current density or by the use of heat or agitation. Identical current distribution can be obtained by changing the position of the anodes and using shadows to compensate for the change in anode position.

Normally, a bath is run in such a fashion that changes in the solution and in the anodes are immediately reflected by a change in the deposit. In this respect, it can be said that under normal operation, the cathode is dependent on solution changes and on proper anode control. However, it is good to know that the cathode may be treated almost independently of the bath and the anodes.

To illustrate the preceding points, let us assume that a Rochelle copper bath is normally operated at 140°F. However, in order to solve a racking problem, a wax is used that softens above 125°F. Now, by decreasing the free cyanide and increasing the current density, it may be possible to obtain the same quality deposit at 120°F. New limits are set on the bath and the same result is obtained. The new limits may not be as desirable as the old ones because if the bath is operated at lower temperatures and a lower free cyanide content, the anode solubility may not be as satisfactory. Nevertheless, the bath will operate perfectly with more frequent chemical adjustment and when the wax racking method is dispensed with, the bath may be returned to the old limits. Almost any bath may be temporarily changed in a similar fashion.

If a nickel bath is being used for plating on steel, it may be desirable to operate the bath at low pH since the pH is more easily controlled in the low range. If the nickel bath is to be used temporarily to plate zinc-base die castings, it may be desirable to operate at high pH in order to reduce the rate of attack of the zinc alloy and thus reduce the rate of contamination of the nickel bath. Nickel carbonate may be added to the bath to increase the pH and by changes in the boric acid content and current density, a satisfactory plate can be obtained.

These examples are hypothetical and should not be used as a guide. In addition, baths should not be changed without experimentation to determine what compensating changes are required in other chemicals, in current density, temperature or agitation. The greatest danger is that the plating range will be reduced. But if plating-range tests are made and the changes in the bath are not radical, changes may be made to meet special plating requirements. Any experimentation should be preceded by a study of the chemistry of the bath to determine the function of each bath ingredient and how compensating changes can be made. Also, it will be found that some changes are impossible. If it is desired to operate an alkaline tin bath at low temperature, it will be found that the cathode efficiency drops rapidly as the temperature is dropped and that there are no adjustments that can be made to make the bath plate tin at room temperature.

Properties of the Deposit

The science of electrodeposition has been looked on as a division of electrochemistry and a knowledge of chemistry is really essential to bath formulation. However, the product is a metal that belongs to the realm of the metallurgist.

The conditions under which metals are produced by electrodeposition are very different from casting methods, but the deposit is, of course, governed by the laws for the metals. The fundamental difference between plating and casting is that in the first process, the metal is formed very slowly and it is formed cold as compared to the casting process. In general, the grain size is very much smaller than that in cast metals and it is to be expected that the deposit will have the properties associated with small grain size. The grain size is often so small that it cannot be resolved under the microscope. On account of this small grain size, electrodeposits are similar to cast metals after extreme cold working. The deposits will, in many cases, be hard and brittle due to the fine grain size, but, of course, these properties can be changed by heat treatment. Also, by changing the bath composition, deposits can be formed that are soft and ductile.

For each application, certain physical properties of the deposit are required. Before measurements are made, it is well to consider the mechanism of formation of the deposits for their properties will be governed by this mechanism.

The standard methods of measurement are not readily

adaptable to the evaluation of the deposit. Methods should be used that can be applied to thin sheets for the deposit is usually quite thin and the properties of the thin deposit may be different from those of a heavy section that was produced merely for testing. The first layers of metal that are laid down are often finer-grained than later layers so that the tensile strength may decrease as the deposit become thicker. In order to obtain a specimen, a parting compound, such as graphite or a very thin layer of grease, may be applied to the cathode. Or, if the cathode is buffed and cleaned with solvent or magnesia but not etched, it is often possible to strip the deposit from the cathode. However, a deposit that is formed over a buffed cathode is liable to have a finer grain than one that is formed by production methods. Metallographic examination will reveal how the grain size of deposited metal is influenced by the basis metal. If acid copper is deposited on clean etched copper, it will tend to reproduce the grain size of the underlying cast copper. If, however, acid copper is deposited on buffed copper, or any buffed metal, even though the metal is etched, it will tend to reproduce the fine grain of the cold-worked surface. As the plate grows, the grain size will change until a thick deposit will produce grains that are characteristic of the type of bath regardless of the influence of the first layers of metal.

Because of the manner in which the electrodeposit grows, hardness measurements cannot be made with the ordinary equipment. Hardness should be measured by micro methods, either as scratch hardness or with an instrument that makes a very small indentation, such as the Knoop indenter.

Electrodeposits compare roughly with the fine-grained cast metals. If the process is looked on as cold casting, it may then be expected that many properties of the deposit will not correspond to those of a metal formed from a melt and subjected to hot working.

Crystal Structure

The crystal structure of cast metals is best observed by etching. Examination of the cast surface is not very revealing because surface effects due to oxidation and freezing of the surface obscure the manner in which the crystals grew. The electrodeposit can be studied at any point in the process by merely removing the cathode from the bath and examining it. Examination often reveals the crystalline nature of the surface. If the deposit is visibly crystalline, it is known as a "crystalline" deposit. If the deposit has to be examined under the microscope to detect its crystalline nature, then it is "finely crystalline." Surface examination under the microscope may reveal that the deposit is crystalline, nodular, or quite flat. Usually, if the deposits are heavy, they will present a macroscopic appearance which is similar to the appearance of thin deposits under the microscope. Crystalline deposits and even large crystals of metal growing out into the bath are characteristic of simple pure metal salts. Pure silver nitrate, copper sulfate or lead acetate may produce large loose crystals that are of no practical value. However, as soon as secondary salts or foreign substances are introduced into the bath, the simple crystalline structure may be broken up. Sulfuric acid will have such an effect when it is added to copper sulfate so that the deposit becomes finely crystalline. By addition of glue, the deposit may become even more finely crystalline so that it may be sufficiently flat to be bright. All impurities do not produce this effect, for some apparently promote growth perpendicularly to the basis metal and long needles or leaved deposits may form. By experimentation, additives have been developed for many of the baths, which produce a finely crystalline, a smooth, or a bright deposit. The addition agents include a host of organic compounds, colloidal substances and wetting agents, all of which are organic in nature. However, heavy metal salts are sometimes also included as addition agents. An approximate definition of addition agents may be given as follows: Addition agents are substances which improve the quality of the deposit when added in small amounts. Such a definition would include salts of heavy metals which are nobler than the metal of the primary salt. The primary salt, in this case, is the salt supplying the ions of the metal being plated. In alkaline baths, basic salts of the metal may be trapped in the deposit and produce an addition-agent effect. In general, the alkaline baths will produce sound deposits without the aid of addition agents, although additions may be used in order to obtain bright deposits.

In view of the fact that the deposit will trap metals, metal compounds and organic substances, it may consist of a pure metal, solid solutions, intermetallic compounds, two or more metallic phases or of metal and oxides, sulfides or organic substances. Usually the amount of oxygen, sulfur or carbon present, as the elements of an organic or inorganic compound, is very small: of the order of a few hundredths or a few thousandths of a per cent. Therefore, the analysis of the deposit will usually be 99.99 per cent pure metal. If a second metal is present, it may run as high as several per cent before it will cause any noticeable difference in the appearance of the deposit.

The crystal form of the deposit is usually in the crystal system that is normal for the metal. Occasionally, the basis metal influences the deposit to such an extent that an allotropic or such a crystal form is produced that is abnormal for the particular metal. Electrodeposited cobalt has been shown to consist of a mixture of face-centered cubic and hexagonal close-packed forms.[1] Chromium has been shown to be deposited in body-centered cubic and hexagonal close-packed forms.[2]

Recrystallization

Because of the small grain size of electrodeposited metals, recrystallization usually yields small equiaxed grains, but the recrystallized grain size is influenced by the original grain size of the deposit. Occasionally, unusual effects may be produced due to the presence of small amounts of impurities that migrate to grain boundaries on recrystallization.

Stress

Metals can be deposited under extreme stress. Forces can be trapped in the metal that are so severe that they will tear the deposit from the basis metal. Although extreme stress is unusual, very high stress is normal for deposits of chromium or hard nickel. For example, if chromium is deposited on one side of a thin metallic sheet, the sheet will bend. If the chromium is then etched under proper conditions, the stress will be relieved and the sheet will straighten. The amount of stress can be varied by change in bath composition and by change in plating conditions. In popular bath formulations and normal applications, stress causes no trouble, but with thin sheets, occasionally difficulties arise. If a highly stressed deposit is applied to both sides of a thin sheet, the sheet will not bend because the force will be equal on both sides. But the stress can be great enough to break thin sheets, and this effect, known as the "sandwich effect," is sometimes confused with hydrogen embrittlement.

Hydrogen Embrittlement

Atomic hydrogen can cause embrittlement of the deposit or the basis metal. The hydrogen given off at the cathode is in such atomic or "active" form. It is also said that hydrogen is formed *in situ* or *in statu nascendi.*) This may take place in electrolytic cathodic alkali cleaning, in pickling or in plating. The evolved hydrogen can be eliminated in the cleaning and pickling steps by using anodic treatment and the embrittlement in the plating step can be reduced by changing the plating conditions. The embrittlement may also be relieved by heat treatment after plating. This is common practice in hard chromium plating.

Hard chromium is deposited to produce long-wearing surfaces, but the deposit is often so brittle that it will crack on grinding. The brittleness can be relieved by heat treatment after the plating operation. Although hydrogen may have considerable effect on the ductility, it has been reported that it has little effect on the hardness of chromium.

Hardness

The hardness of a deposit may be changed by changing the current density or temperature of a bath. The hardness is also greatly affected by the presence of organic substances, impurities and addition agents. It was shown that more oxygen is present in hard iron than in soft iron deposits.

Hardness ranges shown in Table 3 were reported by P. J. Macnaughton and A. W. Hothersall.[3] The values in this table show that in plating, it is possible to attain hardness values much greater than those obtained in work hardening.

TABLE 3

The Hardness of the Deposit

Metal	*Type Solution*	*Electrodeposit Brinell Number*	*Brinell Hardness* Fully annealed	Work hardened
Chromium	Chromic Acid	400–950	70	
Platinum		606–642	47	97
Rhodium		594–641	101	
Palladium		190–196		
		385–435	49	109
Nickel	Sulfate	125–420	70	300
Nickel	Sulfate with Organic Colloids	Up to 550		
Iron	Sulfate and Chloride	140–350	69	148
Copper	Acid Sulfate	40–62	40	102
	with Colloids	Up to 130		
Copper	Cyanide	130–160		
Silver	Cyanide	60–79	25	68
Cadmium	Cyanide	12–22	20	34
Zinc	Sulfate	40–50	33–40	52
Tin	Stannate	8–9	4–5	

Layered Deposits

Microscopic cross-section examination will often reveal layered deposits. These may be caused by addition agents, by impurities or by metals that do not fit into the crystal lattice. A condition such as this was produced by addition of lead to a cyanide copper bath.[4] The lead will not fit into the copper lattice so that lead ions build up near the cathode and after sufficient copper has deposited to diminish its concentration near the cathode, the lead deposits in a layer. Another copper layer is deposited over the lead layer and the process is repeated. Small amounts of lead brighten the copper deposit, but large amounts cause an inferior plate.

REFERENCES

1. *Trans. Electrochem. Soc.,* **17,** 571 (1921).
2. *Trans. Faraday Soc.,* **31,** 1253 (1935).
3. P. J. Macnaughton and A. W. Hothersall, *Trans. Faraday Soc.,* **31,** 1168 (1935).
4. W. R. Meyer and Arthur Phillips, *Trans. Electrochem. Soc.,* **73,** 377 (1938).

CHAPTER 5

The Preparatory Steps of Plating

Quality plating is primarily dependent on the reaction that takes place at the cathode during deposition. This reaction can only be favorable if the plating bath has been properly prepared and adjusted. But even if conditions are right in the plating bath and at the cathode, there is no guarantee that the deposit will be satisfactory. It must be remembered that the quality of the deposit is also dependent on the condition of the cathode and on the preparatory steps. If the article to be plated is too porous, nothing can be done to produce a sound, continuous deposit, short of remelting the basis metal and making a new article. It is, therefore, necessary to make sure that the article to be plated is of good quality and to use the proper preparatory treatments prior to plating.

The proper preparation for plating a new item starts with the making up of a fresh plating bath. If the bath is not compounded properly, much time will be lost in adjustments before the bath is ready for plating. If the bath is made up properly, it is possible to begin plating immediately without initial chemical analysis and without adjustments of the chemical content. However, most baths operate better if some initial electrolysis is carried out prior to plating of production pieces. An alkaline tin bath may be made up and used immediately, but a nickel bath almost always works better after some initial electrolysis. Other baths, such as cyanide silver, require only filtration prior to use. Still others require filtration, electrolysis and treatment with activated carbon or clay.

Preparation of the Bath

The first step in preparing a new bath consists of thorough cleaning of the empty plating tank. It is far easier to clean the tank when it is empty than to remove the dirt from the solution after the bath is once made up. All solid particles and dirt should first be removed from the tank by sweeping and wiping. After the tank is thus swept out, it should be examined to determine what further cleaning is required. If the tank contains grease, this should be wiped out with a rag containing a suitable solvent. Further cleaning depends on the type of materials of which the tank is constructed. Rubber or plastic tanks require no further cleaning, whereas wooden tanks may require painting or lining with pitch. If the tank is made of steel its surface will be probably covered with a visible oxide layer and will also bear heavy deposits of oxide and slag at the welds. In many cases, the steel will be covered with a tight black oxide of iron that in itself is a good surface. It is difficult, however, to determine by examination whether the oxide will or will not be loosened after the bath is prepared. Therefore, it is best to remove the scale with acid. If means are available to heat the tank, then it is well to completely fill it with 1 per cent sulfuric acid and heat until the scale is removed. If means for heating are not available, the tank may be wiped with 15 per cent hydrochloric acid which will readily remove the scale. Rubber gloves should be worn during this operation and it should be carried out in a well-ventilated room or near a hood so that the acid fumes are carried away. After the tank is pickled, it should be rinsed with water and if it is not to be used immediately, it should be filled with 1 per cent sodium hydroxide or trisodium phosphate solution to prevent excessive rusting.

After the tank is cleaned and rinsed, it may be partially filled with water so that the chemicals may be added. If the tank is partially filled, it will be easier to agitate without spilling solution over the top. For very soluble salts, it may be convenient to fill the tank one-third, but for salts that are not very soluble, such as the total salts added to a high-concentration nickel bath, the tank should be filled about two thirds; but never full.

The order of mixing chemicals is important. If a cyanide copper bath is to be made up, the alkali cyanide must be added before the copper cyanide because the copper cyanide is insoluble in water. The order of addition of the carbonate, caustic and Rochelle salt is relatively unimportant. Since the chemicals are not pure and since the tap water is hard, there will be precipitation of salts in the freshly made bath. In many baths, these salts must be removed by filtration and they are more rapidly removed while the solution is concentrated. Thus, it is often time saving to start filtration while the solution is concentrated. The bath should then be made up to volume after most of the precipitate is removed since more salts may be precipitated on further addition of water.

At times, it is convenient to make concentrates for the purpose of preparing fresh baths and for additions to old baths. The advantages of concentrates are that they can be filtered and are more convenient to add to baths than salts in solid form. The disadvantages are that they require storage space and they increase the volume of the bath to a greater extent than salts (it may be necessary to remove a part of the bath).

When large volumes of solution are made, the salts should be added slowly and the solution should be agitated. Agitation may be provided by hand stirring with a large paddle, but if many solutions are made, a portable electric mixer will be convenient. If the salts are added too rapidly, they may form a cake on the bottom of the tank which will be very difficult to dissolve.

After all the salts are in solution, the specific gravity of the bath should be measured. This is commonly done with a Baumé hydrometer. If the right amounts of salts were used and they were all dissolved, the proper gravity will be obtained. If the gravity of a fresh solution is not known, it can be determined by making up 1 liter of solution and measuring the gravity. The gravity measurement is an immediate check on the over-all salt content.

After the bath is of suitable gravity and filtered, a plating test should be made. This may be done by either plating on a simple flat cathode, plating on an actual article, or best, by running a plating-range test. If plating-range standards are not available, it is well to record a sketch of the appearance of the plate in the log. The plating test will indicate whether the bath may be used as is or further adjustments are necessary. Adjustments of pH or addition agents may be required, or the bath may need initial electrolysis and purification.

It is often convenient to make a complete chemical analysis of a fresh bath, although this is not necessary if a gravity check and a plating test are made. However, a record of gravity, analysis and plating test is valuable for the log until enough data are accumulated to dispense with some of the analyses.

Purification steps are best carried out simultaneously. These consist of filtration, carbon treatment and electrolysis. The activated carbon should be loaded into the filter and the solution pumped over the carbon and through the filter. A large cathode may be suspended in the tank and the bath electrolyzed until the deposit looks good either on the cathode or by a plating test. Electrolysis for purification is most effective at low current density; usually a current density of 5 amperes per square foot is used.

After the bath has been adjusted to chemical limits, purified and tested by plating, it should be tested at full load. The tank load can be calculated from the number of racks that it will hold and the current per rack. A cathode area of approximately full capacity should be used and the proper number of anodes should be placed in the tank. The number of anodes can be calculated from the total current, the area of the anodes and the recommended anode-current density.

The tank is operated at full load in order to check the behavior of the current. The tank voltage is noted and the tank is examined for electric faults and good contact of the racks and the anode hooks at the bus bars. If several tanks are supplied with current from one generator, it is well to check the effect of full load of one tank on the operation of another at full load. An electrolytic cleaner suddenly drawing current may cause considerable drop in tank current if the total current is near the capacity of the generator. This will only be serious if it interferes with plating thickness and the effect can be estimated from the observed current drop.

If a bath is being set up for the first time, it will be profitable to keep good records. If an adjustment of pH is required, record the amount of acid or alkali added and the pH before and after additions. From data of this sort, a graph can be plotted that will show the exact quantity of chemical to be added to bring the pH from any point to within the required pH limits. Useful data can be accumulated in a similar manner for adjustment of surface tension where this is required. If a few gravity readings are taken before and after chemical additions, a chart or graph can be prepared that will indicate the total salt for a gravity reading. A little extra time during the initial preparation of a solution will save time on all future adjustments. Charts and graphs prepared from initial data are always useful. When calculations are made for the amount of salt to be added to the bath, these should be recorded along with the tank volume. Calculations can then be made for the number of pounds of salt to be added to increase the bath concentration 1 ounce per gallon.

After the bath is found suitable for full-scale operation, the geometry of the plating system should be considered. Anodes should be spaced so that the current reaches the racks equally from all directions, unless a special current-distribution problem exists.

The plating racks should be examined to determine whether the solution drains readily and is not trapped in the racks, causing drag-in and rinsing troubles. It is also well to check the rack contacts. This can be done by hanging pieces on the rack and checking with an ammeter across each contact for equal ability to carry current from the rack to each individual piece.

The efficiency of hoods, heaters and means of agitation should be examined.

At the same time as the plating tank is set up, the rest of the plating line should be put into operation. This will include rinsing, cleaning, etching and special steps, such as striking or bright dipping.

Cleaning

The cleaning operation is usually carried out in some type of alkaline cleaner. The type selected depends on the material to be cleaned and on whether the cleaning is to be electrolytic or by soaking. Electrolytic cleaners are widely used because of their rapid, efficient action. The same cleaning action can be obtained from electrolytic cleaners in 15 seconds to 1 minute as in 5 to 20 minutes from soak cleaners.

Cleaners may contain sodium phosphate, sodium hydroxide, sodium silicate, sodium carbonate, sodium borate or any combination of these salts. Most cleaners consist of phosphate or silicate with some hydroxide or carbonate added. In addition, they may contain soap, wetting agents or sodium hexametaphosphate. Satisfactory cleaners may be prepared from these chemicals for almost any cleaning application. There is not sufficient information available about wetting agents and the behavior of these compounds varies widely with each application. Common practice is to consult with detergent company representatives for recommendations on each new application.

The type of cleaner that is used should be known for analytical control purposes. The chemical control methods for a phosphate cleaner are not suited to a silicate cleaner, etc.

When a cleaner is recommended, a sample should be obtained and cleaner tests should be conducted. The simplest sure cleaner test is a full-scale test. If the cleaner will cleanse the articles of dirt and grease and leave the rinse with a continuous film of water over the work, then the cleaner is performing satisfactorily. If several cleaners are to be compared, some preliminary work in the laboratory by a standard cleaner test is recommended.

Minor cleaner troubles may develop. These are usually attributed to compounds, in the cleaners, that do not readily rinse from the work. The simplest way of testing for such compounds is to allow the rinsed work to dry and examine the surface for films. Better rinsing may solve the problem.

In making up a cleaner, the same mixing precautions should be used as previously outlined for a plating bath and an initial gravity check should be made.

Electrolytic cleaners may be operated with the work as the anode or the cathode. Anodic cleaners are often preferred because metal impurities cannot deposit on the work. However, if the work is reactive to anodic current, then cathodic cleaning may be required.

Still or soak cleaners are preferred where small parts are to be cleaned in a basket and current will not reach the inner parts. They are also used where current is not available or where cleaning time is not important. Soaps and wetting agents may be used in still cleaners that otherwise could not be used in electrolytic cleaners due to the electrolytic action causing excessive foam formation.

Degreasing

In certain applications, it is advantageous and sometimes necessary to degrease prior to electrolytic cleaning. If the articles are degreased initially, the cleaning problem is greatly simplified because little or no oil or grease will remain after subjecting surface oils to the action of a solvent. The thin surface film that remains is easily removed in an electrolytic cleaner and the cleaner itself does not become as contaminated as when it is required to perform the whole cleaning job.

Degreasing may be done by dipping the work in a cold solvent or it may be done by exposing the work to hot vapor. The vapor will condense on the metal and only clean solvent will come in contact with the work.

Practically, no combination of oil and grease will withstand the double cleaning step of vapor degreasing and electrolytic alkaline cleaning. However, there are certain oils used in rolling mills that are so changed by the heat and pressure of rolling that they defy the best cleaning methods. If such a resinlike film is encountered, it must be removed mechanically or the trouble must be eliminated at its source by specifying the type of rolling oil used.

Pickling

The preparatory step of every plating process consists of some type of etching which is required to remove inorganic surface films and expose the metal to the action of the plating bath. If this step is not carried out properly, peeling, blistering or poor adhesion of the deposit will result. An etchant must be used that will expose the metal, that will render the metal active and that will not pit or excessively roughen the basis metal.

Etching is usually done with mineral acids, sulfuric being preferred and hydrochloric being used if necessary.

Sulfuric acid is cheap and does not give off irritating fumes as does hydrochloric acid. It is a strong, active acid and may be obtained in high concentrations. In pickling steel, 5 to 10 per cent sulfuric acid is commonly used with a pickling time of approximately 1 minute at temperatures of 140 to 180°F. For metals more active than steel, lower temperatures or lower concentrations may be used. At times, hydrochloric acid is more desirable. In high-carbon steels, hydrochloric acid will not leave as much carbide smut on the surface as sulfuric acid.

Pickling may also be done electrolytically, the advantage being that the solution of metal can be treated by the current and the amount of metal removed can be more closely controlled.

Many acids and many combinations of acids are used in the various plating processes. For instance, both nitric and hydrofluoric acids are used to pickle aluminum and its alloys before plating. Nitric acid in various concentrations is excellent for preparing aluminum surfaces, while the combination of nitric and hydrofluoric acids is commonly used for alloys containing silicon.

Rinsing

The rinsing steps are just as important as the other steps in the plating process. Occasionally, films picked up in the cleaner are difficult to rinse. At times, there are slots, holes or cracks in the article, or between the article and the rack, that are difficult to rinse. Rinsing is generally done in running water, but it may also be done effectively by spray.

A double rinse is often required for satisfactory removal of salts from the surface. A rack that is plunged below the surface of running water, allowed to hang in the air for a few seconds and then plunged again will be more effectively rinsed than if it were suspended in the rinse for the same length of time. If the rack is to be dipped twice, it is better to use two rinse tanks. Some saving in water is possible by allowing the slightly contaminated water from the second rinse to flow through the first rinse as shown in Figure 8.

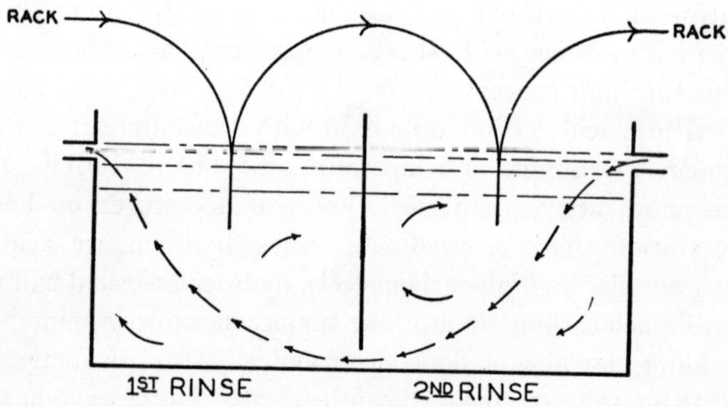

FIGURE 8

Double rinse with underflow

Striking

A very thin layer of metal is often deposited prior to plating in order to obtain bond of the deposit to the basis metal. For instance, in the silver plating of steel, it is impossible to bond the silver no matter what preparatory steps are used. Steel is passive in a silver bath. It can be plated, but it cannot be bonded. If a strike is used prior to silver plating, then bond will be obtained. The strike may be either a cyanide-copper or a cyanide-silver strike, but it must be a bath where considerable gassing takes place along with plating. Striking for approximately 1 minute in a low-efficiency bath is required. Under these conditions, the steel is active and the thin strike deposit will bond the silver to the steel.

There is no accurate definition of a strike, but in cases where a thin bonding deposit is required, the bath is referred to as a strike. A strike bath or a flash bath may serve also other purposes. If such a step is used, it is often possible to get better coverage in baths of poor covering power. Furthermore, the strike may have some cleaning effect since its electrolytic action is similar to that of an electrolytic cleaner. In fact, cleaning action is sometimes improved by adding sodium stannate or sodium cyanide and zinc cyanide to the cleaner.

Treatment after Plating

Drag-out tanks are occasionally used, following the plating tanks, to collect drainings and rinsings from which metal can be reclaimed. The drag-out recovery tank may be followed by a cold rinse and then a hot rinse. From the hot rinse, the article may often be blown dry, or it may enter a drier. Special bright-dip solutions may also be used at this point in the plating line to brighten the deposit, improve the resistance to staining or to prevent streaking or water marks.

CHAPTER 6

Cadmium Cyanide Baths

Cadmium is mainly used to protect steel and, to a lesser extent, copper-base alloys, from corrosion. Cadmium is an active metal which means that it will protect the basis metal at the expense of being used up chemically or electrochemically. Since cadmium is relatively active, its use is restricted to mildly corrosive conditions. In general, it affords the same type and degree of protection as zinc and, therefore, it should only be used where it has an advantage over zinc. It is resistant to alkaline solutions used in washing machinery, and it resists salt and oil better than zinc. Because of this resistance, it is used for marine applications and in plating tools that are used in the presence of oil.

The economy of cadmium should be kept in mind if an application is to lead to wide use of the metal.

Not too many years ago, cadmium was regarded as a nuisance metal that had to be separated from zinc spelter. Cadmium has advantages over zinc for special electroplating applications. The metal became popular, but the supply was limited. The increase in demand resulted in a much greater increase in price than with the abundant metals.

Although the main use of cadmium is for corrosion protection of the basis metal, development work has followed recent trends for fine-grained, bright deposits.

Studies have been made to find brighteners or addition agents and to extend the bright-plating range. Extensive outdoor and indoor exposure tests and salt-spray tests have been carried out to evaluate and classify the coating. These tests and experiments resulted in the patented addition agents that are now available. The tests also resulted in the classification of cadmium as equivalent to zinc except for the special applications mentioned, where it shows superior corrosion resistance.

The following bath formula is recommended:[1]

	Still Plating		*Barrel Plating*	
	g./l.	*oz./gal.*	*g./l.*	*oz./gal.*
Cadmium oxide	23– 39	3 – 5.2	20– 32	2.7– 4.3
As cadmium metal	19– 34	2.5– 4.5	17– 28	2.3– 3.8
Sodium cyanide	86–131	11.5–17.5	66–110	8.8–15
Addition agents	As recommended by the supplier			

Chemistry of Cadmium Solutions

Cadmium oxide is insoluble in water whereas cadmium cyanide has a solubility of about 17 grams per liter as $Cd(CN)_2$.

If sodium cyanide is added to cadmium oxide, cadmium cyanide and sodium hydroxide are formed as follows:

$$CdO + 2NaCN + H_2O = Cd(CN)_2 + 2NaOH \qquad (1)$$

Thus sodium hydroxide will always be present if the bath is made from cadmium oxide and sodium cyanide. Sodium hydroxide is beneficial to the bath and provides good conductivity. The concentration of sodium hydroxide should be maintained within 1.5 to 3.0 ounces per gallon.

Sodium cyanide that is added beyond the amount required to form cadmium cyanide will further combine to form sodium cadmicyanide. The formula of this compound is not definitely known but it has been suggested that the compound $Na_2Cd(CN)_4$ be assumed for control purposes.[2]

The equation for the formation of this compound would be as follows:

$$Cd(CN)_2 + 2NaCN = Na_2Cd(CN)_4 \qquad (2)$$

In the cadmium cyanide bath, sodium cyanide is added in excess of that required to form sodium cadmicyanide. The following data should be known: *

* The definitions of the terms used in connection with cyanide baths are given here in order to avoid misunderstandings which may arise from interpretation of these terms from an angle other than that of analytical control.

TOTAL CYANIDE is the entire amount of cyanide in the bath expressed as sodium cyanide.

COMBINED CYANIDE is the amount of sodium cyanide added to react with cadmium oxide to form sodium cadmicyanide (where the complex $Na_2Cd(CN)_4$ is assumed).

FREE CYANIDE is total sodium cyanide minus combined sodium cyanide.

Free sodium cyanide contributes to the conductivity of the bath and aids in anode corrosion. It should be between 8 and 10 ounces per gallon.

It has been found by experiment that the total concentration of the bath can be varied if the ratio of sodium cyanide to cadmium metal is held within limits.

$$NaCN : Cd = 3.75 \pm 0.4$$

All cyanide baths form sodium carbonate by decomposition of the alkalis in the bath. Usually, the carbonate content is held within limits since the plating characteristics will change for very high or very low carbonate. For a cadmium bath, it is recommended that the carbonate is held within 2 to 6 ounces per gallon. If the carbonate becomes excessive, it may be removed by chilling the bath and freezing out sodium carbonate. It may also be removed by precipitation by a proprietary compound.

Bath Preparation

Cadmium cyanide plating baths may be operated in either steel or rubber-lined tanks. Steel tanks are satisfactory and more readily obtainable than rubber- or plastic-lined tanks, but the insulated tanks are preferable, because they prevent stray currents.

In cyanide baths, the sodium cyanide is dissolved first and then the metal compound is added. When the bath is made fresh, it is best to bring all chemical ingredients up to the high limit, except sodium carbonate which should be brought up to the low limit. Initial analyses and plating tests should be run to determine the original condition of the bath.

Bath Control

After placing the bath in operation, it is controlled by analyses for cadmium, free or total cyanide, sodium hydroxide and sodium carbonate. If addition agents are used, they are controlled by plating—range test or by addition according to a schedule based either on operating time or on the amount of metal deposited.

If drag-out losses are not high, the required additions of cadmium oxide to the bath should be low. If the bath does require considerable amounts of cadmium oxide, it indicates that the anodes are not supplying sufficient metal and the anode area should be increased.

Noble-metal impurities will discolor the deposit. The impurities can be removed by low-current-density electrolysis or by treatment with cadmium sponge followed by filtration.

The deposit may occasionally become rough because of impurities suspended in the bath. The suspended material can be removed by filtration.

The plating rate is checked by thickness tests.

Bath Characteristics

The bath is usually operated at a cathode current density of 20 amperes per square foot which will result in a deposit of about 0.3 mils in 10 minutes. The plating range will be about 5 to 50 amperes per square foot, but higher current densities can also be employed by increasing the metal content.

If high current densities are employed or if the bath tends to heat, as in barrel plating, then cooling coils must be used to keep the temperature below 95°F.

Proper solution of metal from the anodes is obtained at an anode-current density of 20 amperes per square foot.

The cathode efficiency is high, but will vary from 85 to 95

per cent. It can be increased by increasing the metal content and decreased by addition of sodium cyanide.

Both the throwing power and covering power of the bath are high so that plating on complicated pieces is possible without special racking.

Preparation of the Basis Metal

Steel is generally prepared for cadmium plating simply by cleaning, pickling. Electrolytic anodic alkaline cleaning and sulfuric acid are desired in these steps. The cadmium bath itself has some cleaning action since it is quite alkaline and some gassing takes place at the cathode.

If heavy deposits of cadmium are required, it will be found that they are not bonded to the steel and it will be necessary to deposit acid cadmium prior to cyanide cadmium to obtain bond.[3]

References

1. G. Soderberg and L. R. Westbrook, *Trans. Electrochem. Soc.*, **80,** 431 (1941).

2. G. Soderberg, *J. Electrodepositors' Tech. Soc.*, **13,** No. 9 (1937).

3. U. S. Patent 2,222,398.

CHAPTER 7

Chromium Baths

The hardness, tarnish resistance and brilliance of electrodeposited chromium make it suitable for diverse uses. Advantage is taken of its hardness and low wear rate in applications such as tools and moving mechanical parts. Its brilliance and tarnish resistance are contributing factors in its widespread use on automobile trim and plumbing fixtures. Worn parts, built up with chromium, have given up to ten times the wear of the original part. The increase in wear resistance in most of these applications is due to the hardness of the deposit. However, in other applications, such as cutting tools, the low coefficient of friction of the chromium is equally important.

In engineering applications, reference is often made to hard or industrial chromium and also to bright chromium. In ordinary plating practice, hard and bright chromium are one and the same, for bright deposits are hard and, in addition, the control of the hardness of the deposit is too difficult for the average plater.

Actual control may not be difficult if means are available to measure hardness, but the measurement of hardness of thin chromium deposits is too difficult for control purposes. The same bath is used to plate "hard" and "decorative" chromium. In fact, it may be said that there is only one chromium bath. By changing the concentration, the temperature and the current density, the characteristics of the deposit can be changed. The effect of bath temperature on the hardness of the chromium deposit is shown in Figure 9.

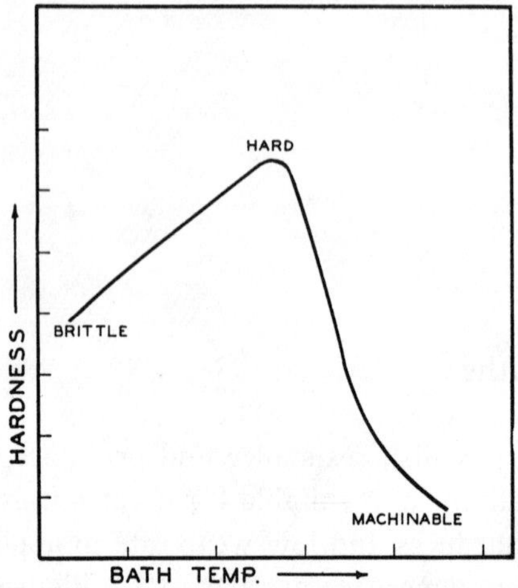

FIGURE 9

Chromium Hardness vs. Bath Temperature

Formulation

The chemistry of a chromium bath is simple. Chromic acid is added to water to supply the chromium metal and a small amount of sulfuric acid is added as a catalyst. The bath does not require "free" chemicals and does not build up in products of electrolysis as chromium is removed. Insoluble anodes are used, chromic acid is added when needed and a definite ratio of CrO_3 to H_2SO_4 is maintained. The baths most commonly used contain either 33 ounces per gallon or 53 ounces per gallon of CrO_3 and the CrO_3 to H_2SO_4 ratio is maintained at approximately 100 to 1.

Operation

The operation of a chromium bath is relatively simple, but as in all baths, there are certain characteristics that must be watched. The solution, in addition to being a plating bath, has fair cleaning action and can also be used to etch steel if the steel is treated anodically in the bath. Not only will it etch steel on anodic treatment, but the etching action is more effective on many alloy steels than with the more common methods of etching. In the common etching methods, using hydrochloric or sulfuric acid, a smut is often left on alloy steels which, if not removed, may cause poor adhesion between the steel and the electrodeposited overlay.

Since the chromium bath has the combined features of cleaning, etching and plating baths, it may be used as a complete plating line with the addition of nothing more than a rinse tank. To illustrate the simplicity of operation, it is well to consider how some manufacturing plants use the bath for salvage of tools. The steel to be plated is immersed in the bath and treated anodically until the desired etch is obtained. The current is then reversed and the steel treated cathodically until the desired thickness of plate is obtained. Chromic acid baths have been used in this manner for limited plating in the vicinity of the plant tool-salvage department. This practice, however, is not recommended for continuous plating or for large-scale plating.

For continuous quality chromium plating, it is necessary to take the bath characteristics into account. The outstanding characteristic of the chromium bath is its limited plating range. The plating range is limited because the covering power and throwing power are low. Also the plating range shifts with change in bath conditions; particularly does the plating range shift with change in temperature or change in current density. However, a shift due to change in temperature can be compensated for by a change in current density. As the temperature is increased the current density must also be increased.

The shift in plating range with temperature change is shown in Figure 10. Since the current does not distribute well over the cathode, it is essential to use an average current density well within the range and to increase the current density as the temperature is increased. It is convenient to express the current density in amperes per square inch since the current density is high by comparison with other still-plating baths.

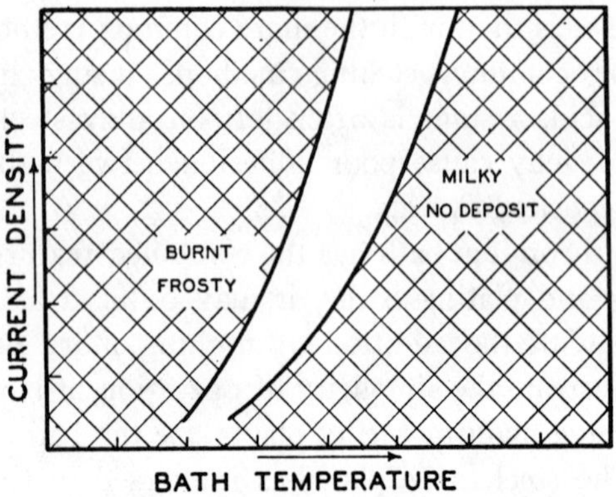

FIGURE 10

Chromium Plating Range; Current Density vs. Bath Temperature

Bath Components

The concentration of sulfuric acid in the chromic acid bath is very low and it only serves as a catalyst. Its effect on the conductivity of the bath is negligible. However, it is very important that the recommended ratio of CrO_3 to H_2SO_4 is maintained. The ratio may vary from 50 : 1 to 250 : 1 but in most cases, satisfactory results are obtained if the ratio is held at approximately 100 : 1. It is imperative that the catalyst ratio is held at an optimum since the plating range of the bath is narrow even under the best conditions.

The chromic acid bath will not plate at all unless the special requirements of catalyst are met. However, if the chromic acid to catalyst ratio is held within proper limits, the behavior of the bath then depends only on the chromic acid. The chromic acid is the source of metal and also provides for the conductivity of the bath, which is good. Since with an optimum concentration of catalyst, there is only one other component present, little can be gained by changing the chemical content of the bath. The dilute bath has a higher current efficiency than the concentrated bath, but a higher voltage is required for its operation.

The difference in power savings between the dilute bath and concentrated bath is small.

Materials of Construction

Chromium-plating tanks may be either lead or brick lined. Antimonial lead or lead-tin-alloy linings are more resistant to corrosion from the bath than chemical lead. The lead or lead-alloy linings are susceptible to the effects of stray currents which may cause chromium to deposit on the lead, may rob the cathode of current, may allow current to flow to areas on the cathode where it is not desired and may accelerate corrosion of the lining. However, if proper precautions are taken, the lead linings are satisfactory and will last for years. Brick lining has the advantage of eliminating stray currents, but it is more difficult and expensive to install than lead.

Bath Preparation

The preparation of the bath is relatively simple since the chromic acid is readily soluble. After the chromic acid is dissolved and the proper amount of sulfuric acid added, the bath is ready for use. Filtration is not required, although it is well to decant the bath occasionally to remove the sludge that accumulates with use.

Table 4 gives the specific gravity readings corresponding to various CrO_3 concentrations.

Anodes

Lead, lead-tin or lead-antimony are used in practice. These lead or lead-alloy anodes are insoluble in the bath. However, during operation, the surface of the anodes becomes covered with a film of lead peroxide which will oxidize any trivalent chromium present.

Iron anodes are occasionally used in applications where the weight or physical characteristics of the lead-alloy anodes are objectionable. The build-up of iron in the bath, however, will eventually necessitate dumping of the bath.

TABLE 4

Specific Gravity and Chromic Acid Content

Specific Gravity at 15°/4°C.	*Baumé at 60°F.*	*CrO_3 oz./gal.*
1.13	16.7	24.8
1.14	17.8	26.8
1.15	18.9	28.8
1.16	20.0	30.6
1.17	21.1	32.6
1.18	22.1	34.4
1.19	23.2	36.4
1.20	24.2	38.6
1.21	25.2	40.3
1.22	26.2	42.3
1.23	27.1	44.2
1.24	28.1	46.2
1.25	29.0	48.2
1.26	29.9	50.2
1.27	30.8	52.2
1.28	31.7	54.5
1.29	32.6	56.5
1.30	33.5	58.7
1.31	34.3	60.7

The presence of trivalent chromium and iron will affect these readings.

Operation of the Chromic Acid Bath

The chromic acid bath is a good example of how a bath is operated by a knowledge of its simple variables. If the following characteristics and variables are kept in mind, it will be found that the optimum operating conditions can be obtained.

1. The bath operates at a high current density.

2. If the current density is increased, the temperature must be also increased.

3. If the current density is increased, the plating rate will be also increased.

4. If the current density is increased, a higher voltage is required.

5. If the plate is bright, it is also hard.
6. The throwing power is poor.

The operation of the bath can be best illustrated by examples.

Let us assume that it is desired to plate 0.1 mil of chromium on steel and that the plating rate is not important. In a 33 ounce per gallon bath, a bright plate will be obtained at 0.75 ampere per square inch and it will take about 12 minutes to deposit 0.1 mil. If the bath has a number of parts in it so that the total cathode area is large, the bath temperature will increase because of the resistance of the electrolyte. If the bath gradually increases in temperature to 105°F., it may be necessary to increase the current to 1 to 1.5 amperes per square inch and to cut the plating time to 8 minutes to maintain a bright plate. This will cause further heating and require further increase in current so that, with continuous operation, the bath may come to equilibrium at approximately 120°F., at a current density of 2 amperes per square inch and a plating time of 6 minutes.

From this example, it can be seen that conditions will have to be continuously changed if appreciable total current is used. It is troublesome to continuously change the current density and the plating time. Therefore, for the above application, it would be more satisfactory to heat the tank to 120°F. before starting operation and to cut the heat input as heat is supplied from the electrolysis. If a large area of work is to be plated, it is necessary to have both heating and cooling coils. Tables are available for the plating rates of chromium baths and they should be used as a guide.[2] However, the tables were determined under conditions of uniform current density and give only approximate values for irregular cathodes. In practical applications, a prepared rack is placed in the chromium-plating tank at the proper temperature and the current density is increased until the plate becomes bright. The total current or voltage is noted and this or some slightly higher value is used for operation. From the temperature and the range of brightness at that temperature, the approximate plating rate is estimated from the tables. The actual plating rate is determined by measurement of the thickness of the deposit.

The major problem in chromium plating is current distribution which requires special attention to racking and to anode positioning in order to compensate for the poor throwing power of the bath. If complicated shapes are plated, a complex anode must be made by bending lead strips or wires to approximately conform to the contours of the cathode.

Preparation of the Basis Metal

For plating on ordinary steel, the common cleaning and pickling processes are satisfactory prior to plating. Alloy steels that are used for tools cannot be properly prepared by this simple procedure, but most alloy steels can be satisfactorily plated if they are anodically treated in a chromic acid bath prior to plating. All alloy steels may not respond to the same treatment although most of them do. For other surfaces, such as nickel- and copper-base alloys, a cleaning, pickling and anodic treatment will have to be determined. Many cathode surfaces do not readily or properly receive a chromium plate unless they are specially prepared, but methods are available for the preparation of most surfaces.[3]

References

1. G. E. Gardam, *J. Electrodepositors' Tech. Soc.*, **20**, 69 (1945).
2. G. Dubpernell, *Trans. Electrochem. Soc.*, **80**, 589 (1941).
3. W. M. Tucker and R. L. Flint, *Trans. Electrochem. Soc.*, **88**, 335 (1945).

Chapter 8

Acid Copper Baths

An acid plating bath such as the acid copper bath is the opposite in many respects to a cyanide bath, such as the cadmium cyanide bath.

A comparison of the general characteristics of the two types of baths is given in Table 5.

Table 5

Comparison of Acid and Cyanide Plating Baths

Characteristic	*Acid Copper*	*Cyanide Cadmium*
Plating rate	almost unlimited	limited
Electrode efficiencies	essentially 100%	variable
Control	simple	somewhat complex
Throwing power	poor	good
Covering power	poor	good
Availability of metal	high	low

The comparison of the two baths, as given in Table 5, will generally hold true for all acid baths and cyanide baths. The main reason for the difference between these two types of baths is that acid baths contain simple ions whereas cyanide baths contain complex ions. More accurately, in acid baths, a fair percentage of the metal is present as simple ions while in cyanide baths only a very small percentage of the metal is present as simple ions and a large percentage as complex ions.

The acid copper bath is most widely used for electroforming, electrotyping and electrolytic refining. It is also used for production of copper powder to be used in powder metallurgy.

Another interesting application is the production of copper screens by use of a continuously rotating drum cathode.[1]

By the electroforming process, it is possible to make sheets, plates, tubes, pipe fittings, bells for musical instruments, tanks, or any shape where a master can be made and an appropriate electrode system can be economically arranged. Thus, the electroforming process should be used for making the article from the pure metal. An outstanding feature of the process is that the reproduction of detail is extremely fine. This is apparent from the fact that acid copper is used in making phonograph records, starting with an original wax master.

The surface of an electroformed copper article can be plated with nickel or other metal or combination of metals, if so desired.

Acid copper is limited in application because of the fact that bright deposits could not be produced by plating alone. Under proper conditions, soft deposits can be produced that may be buffed readily. Recently, several processes have been announced for plating bright copper by the use of proprietary addition agents and as a result, the use of bright copper from an acid bath is expected to increase, even though a deposit of cyanide copper is necessary prior to plating acid copper on steel or other basis metals.

Bath Formula

The following limits in the bath formula have been suggested:[2]

	g./l.	*oz./gal.*
Copper sulfate (crystals)	150–250	20–33
Sulfuric acid	45–100	6–13

The composition of the bath may be varied, depending on the plating rate and the type of deposit desired.

The copper sulfate is the source of the metal ions and a high metal content makes possible a high plating rate. The sulfuric acid is also essential and the solubility of the copper sulfate is limited by the presence of the acid. It is not advisable to operate too near the solubility limit since any increase in salt content or a decrease in bath temperature will result in formation of copper sulfate crystals.

Sulfuric acid provides good bath conductivity and prevents the formation of basic copper compounds. The sulfuric acid also decreases the amount of copper ions present for the same amount of total copper in the electrolyte. As the copper-ion content is decreased, the crystal size of the deposit is also decreased.

In the majority of applications, addition agents are not used but, as previously mentioned, they may become popular for bright deposits. However, they may also be used in electroforming and electrotyping applications. For these applications, it may not be required or even desired that the deposit is bright, and some of the many addition agents investigated may be used, such as phenol, glue, dextrose and many others. The addition agents will decrease the grain size, decrease the tendency to tree and allow higher current densities to be used.

Bath Preparation

The bath is made up in rubber-lined or ceramic tanks. Special precautions regarding the addition of chemicals are not too important except that the sulfuric acid should be added to a large volume of water in order to prevent spattering and overheating. The copper sulfate crystal contains five molecules of water of crystallization and has the formula $CuSO_4.5H_2O$. There is no danger of caking of the crystal, although considerable agitation may be required to bring about solution.

The control of the bath is simple. The gravity is measured, and a titration is made for the acid content. The gravity measurement determines the total content of copper sulfate crystals and sulfuric acid, while the acid content is calculated from the titration. The copper sulfate crystal content is then determined by subtraction of the acid content from the total.[3]

During operation of the bath, the copper sulfate content increases and the sulfuric acid content decreases. The anode efficiency is really greater than 100 per cent, calculated at a copper valence of two, because part of the copper goes into solution at a valence of one. If drag-out is high, the accumulation of copper may be partially taken care of, but some lead anodes may be introduced into the tank to reach a solution balance. Whether the solution is balanced or not, some sulfuric acid will have to be added at intervals to compensate for the drag-out of sulfate content.

Suspended particles in the bath from any cause, including dust from the air, can lead to rough deposits. The suspended particles can be removed by filtration.

By agitating the bath and by increasing the temperature, higher current densities may be used with a resulting increase in the plating rate.

Relatively soft deposits can be obtained if these are required by operations following the plating. However, the conditions that produce soft deposits usually favor crystalline deposits and treeing in high-current-density areas. Harder and finely crystalline deposits are favored by the following conditions.

1. Low copper concentration
2. High acid concentration
3. High current density within the plating range
4. Low solution temperature
5. Low agitation rate
6. Use of addition agents

The first five conditions are those that do not favor replenishment of copper ions to the cathode in the immediate vicinity of the cathode. The sixth condition covers those addition agents that interfere with crystal growth and produce fine crystals.

If rapid plating is desired, those conditions should be used that favor rapid replenishment of the copper ions in the immediate vicinity of the cathode.

Characteristics of the Bath

Since the cathode efficiency is 100 per cent and as cathode polarization in the bath is low, the acid copper bath has a narrow plating range. This means that both covering power and throwing power are low and that particular attention should be paid to the positioning of the anodes. It also means that, unless addition agents are used, treeing will take place in high current density areas and that even if addition agents are used, the deposit will be heavy in high current-density areas (corners and edges).

The high cathode efficiency and the fact that the limiting current density can be greatly increased by the use of agitation make it possible to obtain very high plating rates. However, heavy deposits will become rough after depositing a few mils unless addition agents are used. If it is permissible to remove part of the deposit mechanically, heavy deposits may be produced that will be sound, but they will be too rough to be made smooth by buffing.

Anodes

Rolled copper anodes are preferred for acid copper baths, but cast copper anodes may also be used in applications where smooth deposits are not important.

Because of the tendency of the copper to go into solution faster than it is deposited, the anode area should be kept low. However, enough anodes should be used to properly distribute the current to the cathodes.

The shape and size of the anodes is important. They should be somewhat shorter than the depth to which the cathode is immersed in the tank and they will be more completely used if they are wedge-shaped.

Figure 11 shows steps in corrosion of a regular-shape oval anode and a wedge-shape oval anode.

Preparation of the Cathode

If acid copper is deposited on a noble metal such as silver, all that is required to obtain bond is to clean and etch the basis metal prior to plating. The choice of etch depends on the basis metal; a nitric acid etch or anodic treatment in a cyanide solu-

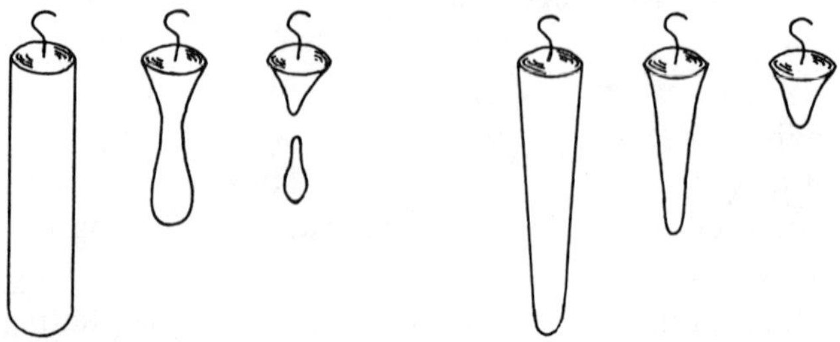

FIGURE 11
Corrosion of Oval and Wedge-Shaped Anodes

tion would provide a satisfactory etch for silver or copper.

If acid copper is deposited on such base metal as nickel, iron or zinc, then bond cannot be obtained directly. Since the copper is noble in acid solutions, it will deposit by chemical displacement and even though the current is flowing, some immersion-plating will take place. The immersion plate will be loose, i.e., nonadherent. It is, therefore, necessary to first plate the base metal with a noble metal using a bath from which the metal will not immersion-plate. A cyanide copper bath is such a bath. The accepted procedure is to plate the base metal with cyanide copper and then deposit acid copper over the cyanide copper. Copper will not immersion-plate on steel from a cyanide bath because steel is a noble metal in a cyanide bath. Therefore, to deposit copper on steel, the usual steps are taken and a cyanide copper-plating step is inserted following the etch and preceding the acid copper plating. When a cyanide copper bath is used in this fashion, it is known as a strike and more will be said about this in the next chapter.

If acid copper is to be used for electroforming, complete lack of bond is essential. The mold may be metallic or nonmetallic. If the mold is metallic, it is cleaned and dipped in a solution of 1 part of light motor oil to 3 parts of naphtha. After evaporation of the naphtha, a light film of oil remains on the mold and this serves as a parting compound over which the copper may be deposited and from which it may be stripped.

If the mold is a nonconductor, such as wax, rubber, plaster, glass or plastic, then a conductive film must be applied. If the surface is porous, it must be sealed with shellac, varnish or lacquer. A conductive film of graphite or bronze powder may then be applied. Or a conductive film of silver may be applied by the reduction process used for silvering glass. This film may be obtained by immersion in a silvering solution or by simultaneously spraying with one solution containing silver nitrate and another solution containing the reducing agent.

Effect of Basis Metal

Occasionally, it will be worth our while to consider some of the experimental work that has been done with the various baths that helps to clarify the mechanism of electrodeposition.

There has been some interesting work done on the study of the initial crystalline form of acid copper deposits on basis metals prepared by various pretreatments.[4] These studies are a key to the adherence of the deposit.

In a hot-dip process, there is some alloying of the basis metal with the overlay which accounts in part for the bond. None of this alloying takes place in an electrolytic process and yet measurements have shown the bond strength to be greater than the tensile strength of the weaker of the two metals concerned. The bond then must depend on the atomic forces between the deposit and the basis metal and the bond strength must be of the order of the strength of a sound metal.

If acid copper is deposited on clean and etched copper, the deposit will tend to repeat the structure of the basis metal at and near the bond line. This indicates that the deposit must be close enough to the basis metal to be influenced by the atomic forces. However, if the basis metal is buffed, it will not influence the crystalline form of the deposit in the same manner.

The crystal structure of a copper deposit may even be influenced by the crystal structure of a basis metal other than copper if the crystal system of the basis metal is similar to that of copper.

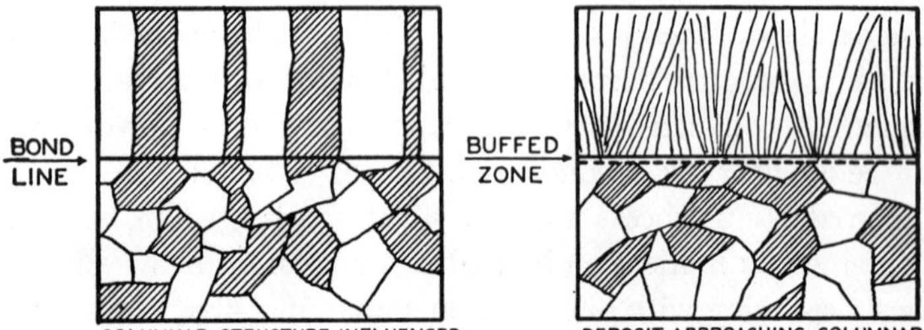

FIGURE 12

Effect of Basis Metal on the Structure of the Deposit

Figure 12 is a diagrammatic representation of the influence of etched basis metal on the deposit.

REFERENCES

1. Shakespeare, *Trans. Am. Inst. Mining Met. Engrs.,* **106,** 441 (1934).
2. J. H. Winkler, *Trans. Electrochem. Soc.,* **80,** 523 (1941).
3. *Natl. Bur. Standards Sci. Paper* No. 275.
4. W. Blum and H. S. Rawdon, *Trans. Electrochem. Soc.,* **44,** 305 (1935). A. W. Hothersall, *Trans. Faraday Soc.,* **31,** 1242 (1935).

CHAPTER 9

Copper Cyanide Baths

Copper is deposited from cyanide baths to prevent carburization of selected areas of ferrous metals. It is also used for part of the specified thickness where more than one metal is deposited and in a number of minor applications. The copper cyanide bath is important to the plater especially for its value as a strike; in this application, a thin layer of copper will make adherence of other metals possible.

The copper cyanide baths may be divided into three types.

1. Plain copper cyanide baths
2. Rochelle copper cyanide baths
3. High-efficiency copper cyanide baths

Each of the types has its use and all contain copper cyanide and sodium or potassium cyanide.

Chemistry of the Baths

The chemistry of the copper cyanide baths is not simple, for more than one complex is formed. However, for bath formulation and control purposes it is safe to assume the following reaction:

$$2NaCN + CuCN = Na_2Cu(CN)_3$$

Thus, it is assumed that two moles of sodium cyanide combine with one mole of copper cyanide; and it will be found on titration for free cyanide that all of the sodium cyanide except two moles will combine with silver nitrate used for the determination. Since this method is used for control and since it measures all the sodium cyanide present but the two combined moles, the assumption is valid on a control basis.

The behavior of the cyanide baths is understandable if the effect of free cyanide is considered (all the sodium cyanide in excess of two moles). When the free cyanide content is zero, a substantial cathode efficiency is obtained even at low temperatures. However, as free cyanide is added, the cathode efficiency drops and plating practically stops at four moles of total sodium cyanide for one mole of copper cyanide (or at two moles of free sodium cyanide).

The plating rate can be increased by increasing the temperature. But there is more required of a plating bath than a substantial plating rate. The deposit must be satisfactory and the anodes must be soluble.

When free cyanide is low, the anodes polarize. A black film is formed and the anodes give off oxygen gas and do not supply copper to the solution. During this process, copper is removed from the solution (at the cathode) and free cyanide builds up with a resulting decrease in the plating rate. Eventually, the increase in free cyanide will aid anode corrosion, but control under such conditions is difficult. Because of these difficult control conditions, the simple low-temperature baths are not widely used. However, they may be used as a strike where plating is not continuous.

For volume production plating, the Rochelle copper bath is generally used. Here, the higher temperatures and Rochelle salt aid anode corrosion and make a higher plating rate possible.

For more rapid plating, the high-efficiency copper baths are used in which the temperature is high, the metal content is high and the free cyanide is low. These baths require several proprietary addition agents for successful operation.

Bath Formulation

The plain cyanide bath may be controlled by adjusting the free cyanide to obtain the desired plating rate. Figure 13 illustrates how the cathode efficiency is affected by the free cyanide content.

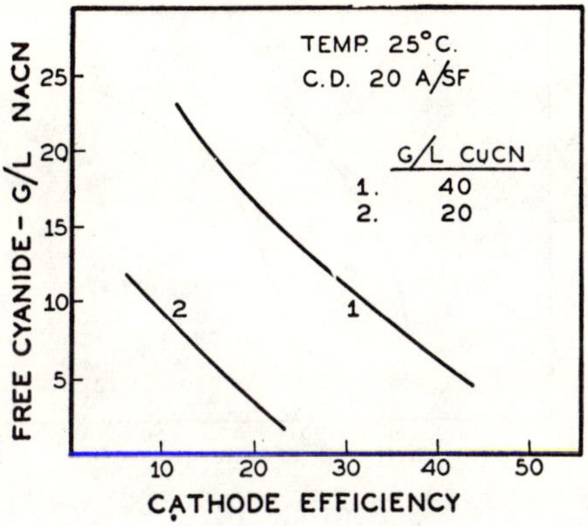

FIGURE 13

Copper Cyanide Cathode Efficiency vs. Free Cyanide

As the free cyanide is increased, the plating rate drops. If copper cyanide is added to the bath, the metal content will be increased and at the same time, the free cyanide will be decreased. Both of these changes will result in an increase in the plating rate. As a result, control is critical. Also, if soluble anodes are used, they should be adjusted so that balance is maintained. If the metal builds up in the bath too rapidly, steel anodes may be used for part of the anode area.

The plating rate may be increased by increasing the current density, but even though the plating rate increases, the cathode efficiency drops rapidly and at high current density, very little is gained. Also, if high current density is necessary, a large anode area may be required to prevent polarization of the anodes.

It is best to increase the temperature if it is desirable to increase the plating rate. Increase in temperature increases the cathode efficiency and also reduces anode polarization. Figure 14 shows how the cathode efficiency increases with increase in temperature.

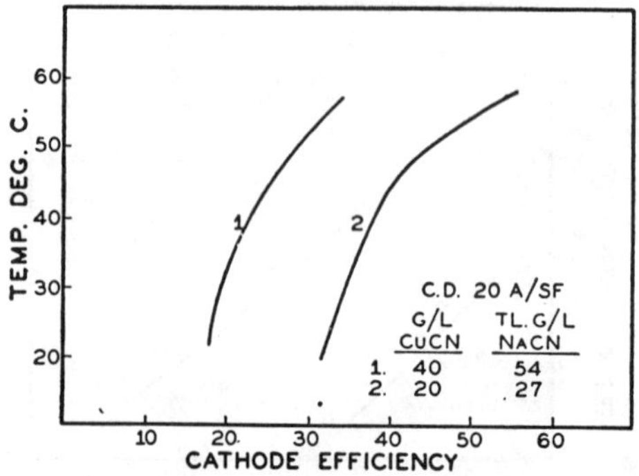

FIGURE 14

Copper Cyanide Cathode Efficiency vs. Temperature

If the plain copper cyanide bath is to be used to produce thin deposits, it is well to measure the cathode efficiency frequently and to balance the bath by adjusting the anode area, the temperature and the current density and by the use of some steel anodes.

If a plain copper cyanide bath is to be used as a strike, it is well to make occasional measurements of cathode efficiency. The bath may be varied widely to meet different requirements. The anode efficiency will not be as important as for continuous plating, because the strike will only be used at intervals. Such a bath may even be controlled by the use of steel anodes and the addition of chemicals to maintain the plating rate.

Very thin deposits of the order of 0.01 to 0.02 mil are often all that are required for striking and this thickness is easily obtained with a striking time of 1 minute in cold baths. However, it is sometimes desired that the strike also has some cleaning power. In this case, it is desirable to decrease the cathode efficiency by increasing the free cyanide, increasing the current density and raising the pH, all of which increase the cleaning power of the bath. It is also desirable to operate at elevated temperatures. High current density, high temperatures and high pH cause the strike to act as an electrolytic cleaner. The pH may be increased by adding sodium hydroxide or trisodium phosphate.

The following formula is recommended for a plain copper cyanide bath: [1]

	g./l.	*oz./gal.*
Copper Cyanide	22.5	3.0
Sodium Cyanide	34	4.5
Sodium Carbonate	15	2.0
Temperature	75–100°F.	
Current Density	2.5–15 amp./sq. ft.	
Anodes	Rolled annealed copper	
Free Cyanide	1.2 oz./gal.	

The Rochelle Bath

The following formula is recommended for the Rochelle bath: [2]

	g./l.	*oz./gal.*
Copper Cyanide	19–45	2.5–6
Sodium Cyanide	26–53	3.5–7
Sodium Carbonate	15–60	2 –8
Rochelle Salt	30–60	4 –8
Free Sodium Cyanide	15–30	2 –4
Current Density	20–70 amp./sq. ft.	
Temperature	130–160°F.	
pH	12.2–12.8	
Anode Current Density	10–30 amp./sq. ft.	
Cathode Efficiency	30–60%	

The Rochelle salt in the bath aids anode corrosion and produces a finer-grained structure than the plain copper cyanide bath.

The cathode efficiency responds to temperature, current density and free cyanide effects in a manner similar to the plain cyanide bath. The pH is held within limits to control the appearance of the deposit. However, changes do not take place rapidly in the bath and with some adjustments, a balanced bath can be obtained.

High-Efficiency Copper Cyanide Baths

The high-efficiency copper cyanide baths carry to an extreme the conditions that will produce high cathode efficiency. These conditions are:

1. High metal content
2. Low free cyanide
3. High temperature
4. Agitation

In addition, the bath is operated at a relatively high caustic content and contains proprietary addition agents.

The following bath limits are recommended for the sodium bath:[3]

	g./l.	*oz./gal.*
Copper Cyanide	90–150	12–20
Sodium Cyanide	100–170	13.5–22.5
Free Sodium Cyanide	3.75–11.25	0.5–1.5
Sodium Hydroxide	22.5–37.5	3–5
Brightener	11.1–18.7	1.5–2.5
Antipit Agent	1.2–1.8	0.15–0.25
Current Density	10–100 amp./sq. ft.	
Temperature	168–185°F.	
Cathode Agitation	2–15 ft./min.	
Anode Current Density	5–30 amp./sq. ft.	

The anode and cathode efficiencies in this bath are both essentially 100 per cent. Thus a balanced bath is easily maintained. In addition, the plating rate is high and the bath is relatively stable, considering the high temperatures employed. The bath gradually builds up in carbonate due to chemical decomposition, but the carbonate can be more easily removed by crystallization from this type of bath than from the baths of lower metal content. If a steel tank is used, the bath may also build up in ferrocyanide, but this does no harm and may be removed by crystallization along with the carbonate.

The disadvantages of the bath are:

1. A strike must be used prior to plating to obtain bond.

2. Cathode agitation is essential to maintain the high plating rate.
3. Addition-agent control is required.
4. Good rinsing is required.
5. Frequent or constant filtration is recommended.

A high-efficiency bath may also be prepared using potassium salts. The potassium bath is similar to the sodium bath except that less metal is required to obtain a high cathode efficiency. Carbonate cannot be crystallized from the potassium bath, but a greater amount can be tolerated and it may occur that the decomposition rate of the bath will be low enough so that removal of carbonate will be taken care of by drag-out.

If the high-efficiency bath is properly controlled, semibright deposits will be obtained that are a good base for deposition of bright nickel.

Contaminants

Cyanide baths may become contaminated with foreign metals or organic materials. These contaminants can be removed by the use of low-current-density electrolysis and treatment with activated carbon. Small-scale tests will indicate the proper treatment.

Bath Preparation

In preparing a cyanide copper bath, the sodium cyanide should be dissolved first. The copper cyanide will then be soluble in the sodium cyanide solution. After the cyanides are dissolved, the other chemicals may be added. Steel tanks may be used, but rubber- or plastic-lined tanks are preferred to prevent the formation of ferrocyanide and to prevent stray currents.

Operation and Control

In any of the cyanide baths, except strike baths that are occasionally used, an adjustment of conditions should be made to obtain balance, so that the anodes are not excessively polarized and that metal is entering the solution at the same rate at which it is being removed.

Such a cathode-current density should be used that good deposits can be obtained at higher and lower current densities. In other words, the average cathode-current density should be somewhere in the middle of the plating range.

By comparison of the three types of copper cyanide baths, it will be seen that conditions can be changed so that one of the baths approaches the performance of another. For instance, if the Rochelle bath is operated at room temperature, its characteristics will be much like that of the plain bath. On the other hand, if it is operated at high temperature and low free cyanide content, it will approach the performance of the high-efficiency bath. Thus a quality deposit can be obtained over a wide range of temperatures and with a wide variation in plating rate. However, the quality of the deposit is not the only measure of a good plating bath. If conditions are greatly different from those recommended, trouble may develop with bath balance and with anode corrosion. If difficulty is experienced, and if bath decomposition is high, frequent additions of chemicals may be required. If excessive additions are required, bath control will be too difficult for continuous plating. In experimenting with variations of a formula, a good log will be helpful to anticipate whether continuous control is feasible. If this is done, there is no reason why experimentation cannot be carried out in full-scale plating.

In all of the baths, analysis for copper and free cyanide is essential for control of the metal content and the plating rate. The pH is controlled in the Rochelle bath, but at the higher pH of the high-efficiency bath, an analysis is made for sodium hydroxide. Also in the high-efficiency bath, the free cyanide content must be determined cold by the use of cracked ice (at 4°C. or less). Addition agents are controlled by plating-range or by special tests recommended by the supplier.

Characteristics

One characteristic of the copper cyanide baths is that heavy deposits cannot be obtained. Good deposits are possible up to 2 mils, but beyond this thickness the deposit becomes rough. Next to limiting thickness, the baths are notable for the drop in cathode efficiency with increase in free cyanide. Cathode efficiency can be increased by the use of low current density and high temperature. Agitation will increase the limiting current density for high-efficiency baths, but if agitation is employed, frequent filtration is required. When a bath is agitated, suspended particles come in contact with the cathode more easily and cause rough deposits. Rolled and annealed anodes generally work well in the copper cyanide baths, although electrolytic anodes are recommended for the high-efficiency baths. If the free cyanide is low, the temperature is low, or the anode-current density high, the anodes will polarize and lead to increase in free-cyanide content, loss in metal content and acceleration of bath decomposition.

Preparation for Plating

The basis metal is cleaned and etched by the usual procedures prior to plating. Copper may then be deposited on the basis metal and good bond will be obtained. If a good copper deposit is obtained, almost any metal may be deposited over it. This is the manner in which the low-efficiency baths function as a strike and low-efficiency plating is required even prior to deposition from a high-efficiency copper cyanide bath.

Heavy Deposits

Commercial installations have recently been made to increase the limiting thickness of copper cyanide deposits by periodic plating and deplating.

References

1. A. K. Graham, *Plating and Finishing Guidebook,* Metal Industry Publishing Co., New York, 1945, p. 47.

2. A. K. Graham and H. J. Read, *Trans. Electrochem. Soc.*, **80**, 341 (1941).

3. H. L. Benner and C. J. Wernlund, *Trans. Electrochem. Soc.*, **80**, 355 (1941).

Chapter 10

Iron Plating

The tendency of iron to rust and corrode has prevented its use as a protective metal coating. Its uses are confined to heavy deposits, and in recent years there has been considerable interest in iron as a metal to use for the build-up of worn tools. This application is important to machine shops where a worn machine part can be brought back to original or oversize dimensions at lower cost than that of making a new part.

The future use of electroplated iron will depend on the application of the known properties of the metal to the greatest advantage. When iron is pure, it has the properties associated with purity, i.e., it can be rendered soft and ductile on annealing. Also the magnetic properties of pure iron can be used.

On the other hand, the metal may be hardened by the ordinary processes of cyaniding, nitriding or carburizing.

The Iron Chloride Bath

In rapid plating, where a hot and corrosive bath may be tolerated, the chloride bath is often used. Thick deposits can be obtained if care is exercised in the control of the bath. Kasper [1] investigated the conditions and prescribed limitations for rapid plating. Later, Stoddard [2] showed that a finer-grained deposit could be obtained by the addition of manganese chloride to the bath. The following composition is typical:

	g./l.	*oz./gal.*
Ferrous Chloride	300	40
Manganese Chloride	5	0.67
Temperature	160–220°F.	
Current Density	50–75 amp./sq.ft.	
Anodes	ingot iron	
pH	1.5–2.0	

The bath can be easily prepared from readily soluble ferrous chloride. Commercial ferrous chloride, however, contains some ferric salts that are detrimental to the bath. In the presence of ferric iron, a rough deposit will be obtained that is unsuitable for any iron-plating application. By working the bath in the presence of a small excess of hydrochloric acid, the ferric iron can be reduced to ferrous iron. If too much acid is present, a deposit cannot be obtained on electrolysis. But if high purity iron is hung in the bath, the acid will have a reducing action on the iron to eliminate the ferric ion. Complete reduction can be recognized by the change in color of the bath from a brownish green to a pure green. After the bath is reduced and worked, the pH is adjusted and the bath is ready for operation. High temperatures must be used since the deposit becomes brittle with decrease in temperature.

Rubber- or plastic-lined tanks or ceramic-lined tanks must be used because this high-temperature, low-pH bath is extremely corrosive to metals.

The pH is the critical control factor in the iron chloride bath. If acid is not added frequently, the pII will risc and a brittle deposit will be obtained. However, if too much acid is added, the cathode efficiency will decrease rapidly; but excess acid is used up by chemical reaction with the iron anodes so that small excesses are rapidly consumed. As the pH rises, the rate of chemical attack drops off very rapidly. This variation in the rate of chemical attack with change in pH makes it practical to add acid continuously. A small amount of acid present at all times, even though the bath is not being used, keeps the iron in the reduced state.

If the bath is operated within the recommended conditions of high temperature and controlled pH, excellent results will be obtained at high cathode efficiency.

High-purity iron anodes must be used. If the carbon content of the iron is much over 0.02 per cent, carbides will be left on the surface. The carbides will be loose, will become suspended in the bath, and will eventually cause a rough plate. In any case, it is recommended that the anodes are bagged in some suitable material.

Through variation of the bath conditions, Stoddard obtained tensile strengths of 50,000 to 110,000 pounds per square inch and elongations of 6 to 50 per cent.

Small amounts of noble metals, such as copper or arsenic, will contaminate the iron bath. These can be removed by electrolysis. Organic impurities can be removed by treatment with activated carbon in the same manner as from acid baths in general.

The Iron Sulfate Bath

Sulfate baths are sometimes preferred for iron plating because they can be operated at lower temperatures and are less corrosive than the chloride baths. Lower-temperature operation greatly reduces the problems of corrosion-resisting heaters, ventilation and evaporation, encountered in a high-temperature, corrosive bath.

MacFadyen [3] recommended the following bath:

	g./l.	*oz./gal.*
$FeSO_4.(NH_4)_2SO_4.6H_2O$	350	47
Current Density	65 amp./sq. ft.	
Temperature	140°F.	
pH	6	

In general, the maintenance of the bath is similar to that of the chloride bath. The bath must be kept reduced by continuous electrolysis or by occasional treatment with iron and excess acid followed by adjustment to the proper pH. A pH of 6 is maintained by keeping a suspension of ferrous carbonate and powdered charcoal in the bath.

This bath is much more subject to pitting than the chloride bath and precipitates form easily at a pH of 6.

The tendency to pit and the tendency to form precipitates can be reduced by operating at higher temperature and at lower pH. Under these conditions, the deposit will be more ductile.

For a ductile deposit and a stable bath, Schaffert and Gouser[4] recommend a sulfate-chloride electrolyte that was particularly developed for electrotyping and surfacing of stereotypes.

	g./l.	*oz./gal.*
$FeSO_4.7H_2O$	250	33
$FeCl_2.4H_2O$	42	5.6
NH_4Cl	20	2.7
Current Density	50–100 amp./sq. ft.	
Temperature	100–105°F.	
pH	4.5–6.0	

Reducing treatment, pH adjustment, filtering and carbon treatment are required during preparation of the bath.

A consideration of various formulas for iron baths will show that iron may be deposited from sulfate or chloride solutions where the acidity is low. Additions of ammonium, sodium, magnesium and other salts may be made that will alter the characteristics of the baths. By changing the pH, the temperature and the current density, the hardness of the deposit can be changed. As the bath composition is changed, the control problem also changes.

The recommended formulas should be used unless a new plating problem arises that warrants considerable experimentation. The low-temperature baths are attractive because the heating and ventilating problems are simple, but the hot chloride bath is easy to operate continuously if the proper pH is maintained and the recommended practice is followed.

References

1. C. Kasper, *J. Research Natl. Bur. Standards,* **18,** 535 (1937).
2. W. B. Stoddard, *Trans. Electrochem. Soc.,* **84,** 305 (1943).
3. W. A. MacFadyen, *Trans. Faraday Soc.,* **45,** 455 (1924).
4. R. M. Schaffert and B. M. Gouser, *Trans. Electrochem. Soc.,* **84,** 319 (1943).

Chapter 11

Lead Plating

The application of lead plating for decorative purposes has met with little favor since the appearance of the coating is not pleasing and it is very susceptible to wear. However, lead baths are used for plating nuts and bolts, storage-battery parts and linings of containers for certain corrosive liquids. They are also used for the production of bearing alloys, for example, by depositing indium over the lead and forming a lead-indium alloy by diffusion.

A lead fluosilicate bath is used for purposes of electrolytic refining, while lead fluoborate and lead sulfamate are used for purposes of electroplating.

For many years, the lead fluoborate bath was prepared by first making fluoboric acid from boric acid and hydrofluoric acid and then adding basic lead carbonate. The preparation of fluoboric acid is troublesome and hazardous due to the rather violent chemical reaction involved and because hydrofluoric acid must be handled with great care to avoid poisonous burns on the skin. In recent years, lead fluoborate concentrate solution and fluoboric acid have been made commercially available and the preparation of the bath has been greatly simplified.

The limits of the plating bath are not critical and the following concentrations are satisfactory:

	g./l.	*oz./gal.*
Lead	80–120	10–16
Free Fluoboric Acid	30–60	4–8
Current Density	10–50 amp./sq. ft.	
Temperature	70–100°F.	

The bath is prepared by adding the proper amount of concentrated lead fluoborate and fluoboric acid to water and adjusting the solution to the proper volume. The solution thus prepared will deposit fair lead, but the deposit will be crystalline, will have a strong tendency to tree at edges, and will have very poor covering power. These conditions are improved by the addition of small amounts of glue or gelatin to the bath. Approximately 0.1 ounce per gallon will be required at the beginning and much smaller additions will be required when the deposit begins to show evidence of crystallinity or treeing. The amount of addition agent can be controlled by a plating-range test.

The concentration of the lead and fluoboric acid is not critical. The lead fluoborate supplies metal and the acid increases the conductivity of the bath. The bath also contains a small amount of boric acid, but this is not controlled. Hydrofluoric acid cannot be present in excess instead of fluoboric acid because free fluoride ion will precipitate lead as lead fluoride.

When glue is added to an aqueous solution, it forms a colloid and it is the colloidal properties of the glue that account for its effect as an addition agent. Colloids carry a charge and because at least part of the colloidal glue is positively charged, it is attracted to the cathode. It is codeposited with the lead. In the deposit, it interferes with crystal growth and aids the formation of a finely crystalline deposit. Only a part of the glue has this function so that the addition cannot be controlled by analysis for total glue content. The glue also undergoes decomposition with time so that the ratio of active to total glue is variable and is not known. If too much glue is introduced into the bath, or if the deposit becomes brittle due to changes in the glue, then it can be removed by treatment with active carbon. The active carbon will not remove all of the glue, but it will remove the portion that is active. If, at times, the bath requires filtration to remove suspended material, some of the active glue will be removed but it cannot be effectively removed by ordinary filtering practice. The colloid will pass through, unless the filter is packed with activated carbon.

Good deposits can be obtained from baths of a wide range of lead concentration, but with higher lead contents, higher current densities may be employed. At a concentration of 7 ounces lead per gallon, good deposits can be obtained at a current density of 10 amperes per square foot. At 10 ounces lead per gallon, an average current density of 20 is used, but current densities of 40 amperes per square foot are possible. If the lead content is increased to 25 ounces per gallon, the current density may be increased to 60 amperes per square foot and with agitation, even higher current densities may be used. At a current density of 60, the bath will deposit 9 mils per hour so that high plating rates are possible.

The bath may be operated with very high or very low acid content, but concentrations of 4 to 8 ounces fluoboric acid per gallon of bath are satisfactory. There is no particular reason for operating with high acid content, and control of low acid contents would be difficult, particularly since pH cannot be controlled by the ordinary electrometric method, therefore, a medium acid concentration is maintained.

Bath Preparation

The bath is made up in rubber- or plastic-lined tanks. Ceramics cannot be used since the bath will attack ceramic ware and form fluosilicic acid. Usually some precipitate is formed when the bath is prepared, such as lead sulfate if sulfate is present in the water. This will settle to the bottom of the tank and cause no trouble, but if a rough deposit is obtained, filtering is indicated. A rubber-lined or stainless-steel filter may be used. The solution may attack stainless steel slightly, but this will be of no consequence for short filtering times.

If dry glue is used, it should be treated with hot water to obtain a dispersion before it is added to the bath. It should then be stirred in the bath for efficient and thorough mixing.

Control

Since the lead content of the bath is not critical, it may be controlled by gravity measurement if a correction is made for the acid. The acid is determined by titration.

Characteristics of the Bath

The anode and cathode efficiencies are essentially 100 per cent so that there is no problem to obtain bath balance. The throwing and covering power is poor, as is typical of the acid baths. If the bath is accepted as a typical acid bath, operation is relatively simple, the only real problem being that of addition-agent control.

Anodes

High-purity lead anodes should be used, although small amounts of the nobler metals that are found in lead will not cause trouble. Antimony and arsenic are not soluble in acids on anodic treatment in the presence of lead. These metals will collect on the surface of the anode as a sludge and eventually drop to the bottom of the tank. If the bath is agitated, the fine metal particles on the anode may become suspended and temporarily cause rough deposits, but the particles will settle and collect with other insoluble material on the bottom of the tank. The bath may occasionally be decanted and the sludge can be removed. It is worth while to remove the sludge occasionally since it is necessary to agitate the tank after additions are made and it is also desirable to agitate the tank occasionally.

If the tank is not agitated occasionally, it will stratify.

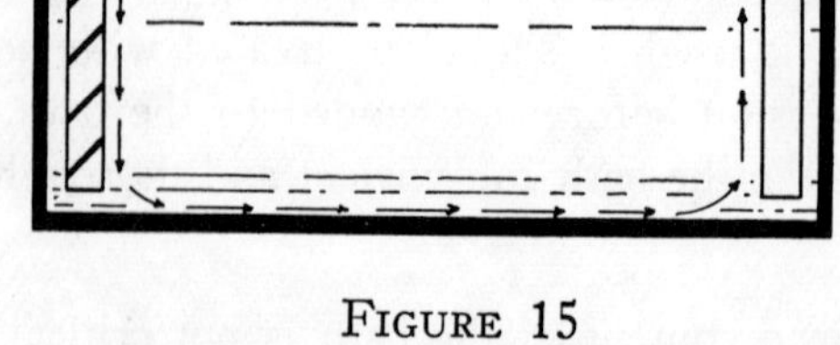

FIGURE 15
Convection-Stratification

When the lead goes into solution from the anode a convection current is set up and the solution of high specific gravity near the anode stratifies along the bottom of the tank. This happens more readily in lead baths than in others because solutions of high specific gravity can be formed with lead salts.

The Cathode and the Deposit

When lead is deposited on steel, the steel is prepared by the usual cleaning and pickling steps. Sulfuric acid may be used for pickling, but good rinsing is important since any sulfates carried into the bath will cause formation of insoluble lead sulfate.

The covering power of the bath may be improved by the use of a copper strike prior to lead plating, but it is of no particular advantage unless a covering power problem exists. With proper cleaning and pickling, satisfactory bond to steel can be obtained without the use of a strike.

In order to protect steel from rusting, a pore-free deposit is necessary. Steel is anodic to lead, and thin deposits are porous. However, a deposit of 1 mil will give good protection unless wear is a factor. If wear is important, deposits of 5 mils thickness may be required.

The Sulfamate Bath

The characteristics and operation of the sulfamate bath are quite similar to those of the fluoborate bath. This bath is operated with several proprietary addition agents and differs from the fluoborate bath in that it is used at a pH relatively high for an acid bath. The reason for this is that sulfamic acid is relatively unstable at low pH and lead compounds are precipitated from the lead sulfamate bath at low pH.

CHAPTER 12

Nickel Plating

The wide use of nickel plating is attested to by the extensive literature on the subject. A great number of patents have been issued and the extent of the literature is indicated by 106 references cited in an excellent paper by Pinner, Soderberg and Baker.[1] If a popular bright nickel bath is operated, it is usually done so under the recommendations and with the addition agents of one of many suppliers. However, for dull nickel plating or for plating for engineering purposes, a formula may be taken from the literature or from a handbook.

Nickel plating is applied primarily for its appearance and for its corrosion protection of steel and of brass- and zinc-base objects. On articles subjected to outdoor exposure, chromium is usually deposited over the nickel, since chromium stands weathering much better than nickel.

Electrodeposited nickel is used for electroforming of sheet, tubes and screen and for build-up of worn steel parts. The composition of the bath used depends greatly on the application, and the characteristics can be changed by shifting the bath limits just as in the copper cyanide baths. However, the formula cannot be changed continuously from one bath composition to another without running into trouble with shift in the plating range, bath control, anode polarization or bath balance.

Bath Formulation

A popular nickel bath is a modification of the original Watts [2] bath that has the following composition:

	g./l.	*oz./gal.*
$NiSO_4.7H_2O$	240	32
$NiCl_2.6H_2O$	45	6
H_3BO_3	30	4
pH	1.5–3.0 or 4.5–5.6	
Temperature	110°F.	
Current Density	40 amp./sq.ft.	

The metal content of this bath is relatively high. The high metal content and high total salt result in relatively good bath conductivity which, in turn, contributes to good metal distribution and wider plating range than a low total-salt concentration.

The presence of chloride ions in the bath promotes anode corrosion and makes it possible to obtain bath balance. The chloride also increases the cathode efficiency at low current density and thus improves metal distribution.

Boric acid serves as a buffer in the bath and thus simplifies pH control. The boric acid is only slightly ionized and furnishes few hydrogen ions, but will furnish them as rapidly as they are removed. Thus it acts as a reservoir for ionizable hydrogen. This function is important because pH control is critical in a nickel bath. It will also be found that a better-quality deposit will be obtained in the presence of boric acid.

Bath Preparation

Bath preparation is carried out in the rubber-lined tanks commonly used for nickel plating. The nickel salts are readily soluble in water, but the boric acid is sometimes troublesome to get into solution. The use of heat and agitation will greatly promote the dissolution of the chemicals. After dissolving the chemicals, the pH is adjusted and the bath is ready for initial plating. It is not likely that the bath will immediately plate satisfactorily. Very often, the deposit will be pitted or off color. If the deposit is satisfactory, as determined by a test panel, then it is often found that bond to the basis metal cannot be obtained. This condition is remedied by working the bath. Electrolysis will remove objectionable impurities and put the bath in operating condition. Electrolysis prior to plating is most effective at low current densities (3 to 7 amperes per square foot). During initial electrolysis, a heavy deposit will be produced and the condition of the deposit will indicate the need for further bath treatment. If the deposit is pitted, a small amount of hydrogen peroxide will usually remedy the trouble. If the deposit is rough or off color, the bath should be treated with active carbon and filtered. It is good practice to work a fresh bath, treat it with active carbon and filter it all at the same time. Small-scale tests can be made to predetermine the amount of treatment required and plating tests can be made on a large bath during treatment to follow the course of the treatment.

Operation and Control

The operating conditions of the nickel bath should be such as to obtain bath balance as closely as possible, with frequent adjustments to hold the bath within the range. If the bath is operated in approximate balance and adjustments are made to keep it in the range, then additions of chemicals will be small. This is important because excessive bath treatment will be avoided. The nickel salts contain impurities that make it necessary to treat a nickel bath before use. If large amounts of nickel salts are added during control, the same treatment will be required. However, small additions of nickel salts will not contain enough impurities to make purification treatment necessary. Small amounts of impurities will be eliminated during normal plating operations, without causing harm.

Usually, a nickel bath is operated according to some accepted formula. The limits on all ingredients will be set as follows: total nickel, sulfate, chloride, ammonia, boric acid and pH. All of these can be controlled by analysis and none of them should change rapidly in the balanced bath, with the exception of the pH. If the bath is operating correctly, the pH should increase and it should frequently be brought to the lower limit by the addition of acid. If initial data are taken on the bath, a graph can be prepared for additions of acid necessary to adjust to the lower pH limit for any pH reading.

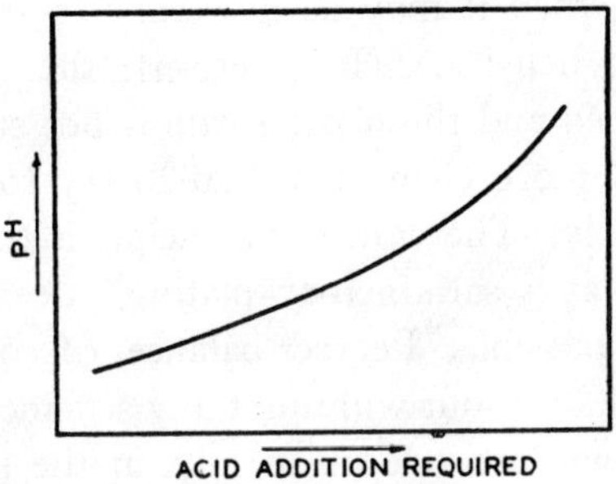

FIGURE 16
Nickel pH vs. Acid Addition Required

It will also be convenient to take gravity readings on the bath since a gravity reading is a rapid check on the over-all solids content. A second operating graph can then be prepared.

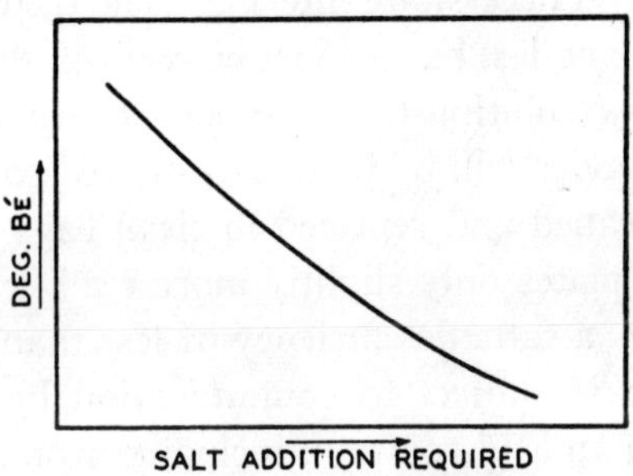

FIGURE 17
Baumé Gravity vs. Salt Addition Required

Bath balance is indicated if changes in composition are small. Changes in gravity, nickel content or rapid changes in pH indicate that the balance is not proper. The balance is obtained by adjustment of the anode area. The cathode efficiency will be fixed by the bath limits, therefore, it will be necessary to adjust the anode area so that the anode efficiency is equal to or slightly greater than the cathode efficiency. If the positioning of the anodes limits the anode area, and thus the anode-current density, then it may be necessary to try changing the cathode-current density. However, if the cathode-current density is dropped and the plating rate is not sufficient, then it is better to add more chemicals than to try to approach bath balance too closely. The working principle is to obtain a satisfactory deposit at a satisfactory plating rate and as close to bath balance as possible. Perfect balance, of course, cannot be obtained because drag-out will upset any scheme for perfection.

Anodes that will corrode uniformly, at the proper rate and without formation of sludge, are not available. But, of course, anodes are available that are satisfactory if they are accepted as not being perfect. Usually the corrosion rate is satisfactory and is accompanied by the formation of an insoluble deposit on the surface of the anodes. This deposit may become suspended in the bath and migrate to the cathode to cause a rough deposit and necessitate filtering. The trouble is overcome by bagging the anodes, but it must be realized that bags restrict replenishment of solution to the anode. If sludge builds up in the bags, the anodes will be further restricted so that they must be removed, cleaned and replaced in clean bags occasionally.

Since nickel plates only slightly more readily than hydrogen (as indicated by a cathode efficiency of less than 100 per cent), the nickel bath is subject to contamination by all the metals that plate from an acid solution, including iron and zinc. Usually the contaminating metals are removed by low-current-density electrolysis. Iron, which is a common contaminant, is removed by oxidation of the bath and increase of the pH with nickel carbonate until the iron is precipitated as ferric hydroxide. The ferric hydroxide is then filtered off.

Organic contaminants are common. These come from the chemicals, tank linings, stop-off lacquers and rack materials. They can cause brittle and cracked deposits and must be removed. Electrolysis may remove organic contaminants, but active carbon is usually more effective. On large installations, it is profitable to continuously filter, treat and electrolyze a nickel bath. This is best done in a tank separate from the plating tank.

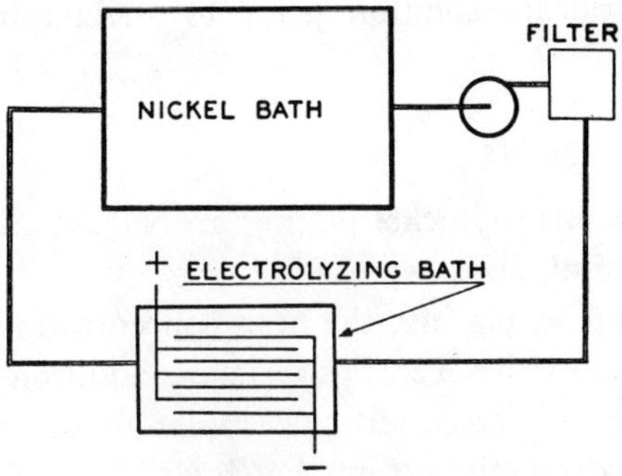

FIGURE 18

Nickel-Electrolyzing Bath

Preparation of the Basis Metal

Steel is prepared for nickel plating by cleaning and pickling prior to plating. Pickling may be done by immersion or by the use of anodic current. In either case, sufficient acid should be present to remove rust or scale if it is present on the steel. The use of inhibitors in the pickling bath is not recommended. Inhibitors are used in pickling steel for economic reasons. The economy consists in saving acid by reducing the attack on the steel while scale is being removed. If steel is removed, of course, acid is used. Heavy scale is hardly ever encountered on work to bc plated. In addition, it will be found that a certain amount of metal must be removed to render the steel active and to secure bond. It is best to remove this quantity of steel by pickling in uninhibited acid or by the faster anodic pickling treatment. In addition, inhibitors may cause trouble in the plating operation if they are not completely removed by the rinse.

If nickel is to be plated on zinc or aluminum, it is best to preplate these basis metals with copper before the nickel coating is applied. These metals are too active to be placed directly in an acid nickel bath. In the case of aluminum, it has been found necessary to either anodize or to immersion-plate zinc from a zincate solution prior to striking with cyanide copper.

Types of Nickel Baths

The applications of nickel plating are varied and, therefore, many bath formulations are used.

For bright nickel plating, the high concentration Watts bath is generally used with several proprietary addition agents. The bath is operated under conditions similar to the conditions of the regular Watts bath, except that hydrogen peroxide cannot be added and the bath cannot be treated with active carbon without destroying the addition agent. Therefore, if purification is required, the bath will have to be taken out of operation, treated and adjusted with fresh addition agent.

A nickel bath may be used to plate directly on zinc-base alloys if the concentration of nickel ions is reduced. The effective concentration may be suppressed by the reduction of the ionization of nickel salt, by means of the addition of various salts. The following solution is an example of a bath recommended for plating on zinc base: [3]

	g./l.	*oz./gal.*
$NiSO_4.7H_2O$	70	9
NH_4Cl	15	2
$Na_2SO_4.10H_2O$	160	22
H_3BO_3	15	2
pH (electrometric)	5.5	

With such a bath, it is not necessary to use a copper strike in order to plate on zinc-base alloys. In cases where thin deposits and such baths are desired that can be operated without close control, a double nickel salt bath of the following composition may be used:

	g./l.	*oz./gal.*
$Ni(NH_4)_2(SO_4)_2.6H_2O$	105	14
NH_4Cl	15	2
H_3BO_3	15	2
pH (electrometric)	5.3	

The double nickel salt bath may be used for volume production of parts where a thin deposit is required. If the parts are small, the plating can be carried out more economically in barrels. The appearance of the deposit may be improved by adding brighteners to the bath. If brighteners are added to a nickel bath of a formula other than that recommended to be used with the brighteners, some experimentation must be done. A minimum amount of brightener should be added and additions should be increased in small steps.

Plating-range tests should be run to establish the optimum bright-plating range. The bath can then be controlled by the plating-range test from the original plating-range data.

Hard nickel baths are used for electroforming and salvage of worn parts. Wesley and Roehl [4] suggest the following bath for depositing thick nickel deposits with a hardness of 380 to 480 Vickers:

	g./l.	*oz./gal.*
$NiSO_4.7H_2O$	180	24
NH_4Cl	25	3.3
H_3BO_3	30	4
Current density	50 amp./sq.ft.	

A tensile strength of 140,000 to 160,000 pounds per square inch was obtained on specimens deposited from this bath with an elongation of 4 to 9 per cent for a 2-inch gage length. Thus the deposit was hard without being excessively brittle. Brittle deposits are occasionally obtained in nickel plating because of internal stress. This internal stress is often so great that the deposit will be torn from the basis metal during deposition. Brittleness and hardness are not necessarily related. The physical properties of the deposit depend on the structure. Banded structures are easily identified metallographically and are the ones that often give trouble.

Earlier baths developed by Wesley contained a high concentration of nickel chloride and no nickel sulfate. These baths would plate nickel of fair hardness and with good anode corrosion because of the high chloride content. The deposits, however, were often of a dark, matte appearance and after looking at ordinary nickel deposits, the appearance of this type of deposit is not pleasing. This point is worth mentioning because if one is interested in plating for engineering purposes, it is well not to be too much influenced by the appearance of the deposit. Microscopic examination will reveal a sound, nodular surface that traps the light and results in a dark-gray appearance. Metallographic examination of a cross-section is a better key than the appearance of a plated surface.

References

1. W. L. Pinner, G. Soderberg and E. M. Baker, *Trans. Electrochem. Soc.*, **80**, 539 (1941).
2. O. P. Watts, *Trans. Electrochem. Soc.*, **92**, 395 (1916).
3. M. R. Thompson, *Trans. Electrochem. Soc.*, **47**, 163 (1925).
4. W. A. Wesley and E. J. Roehl, *Trans. Electrochem. Soc.*, **82**, 37 (1942).

Chapter 13

Silver Plating

Silver plating is used mostly for decorative purposes, but it has also industrial and engineering applications. It is used on electronic equipment, for electrical contacts and for aircraft bearings.

The popular silver cyanide baths contain free cyanide and carbonate in addition to silver cyanide. The following formula is typical: [1]

	g./l.	*oz./gal.*
Silver Cyanide	25–33	3.3– 4.4
Potassium Cyanide	30–45	4 – 6
Potassium Carbonate	30–90	4 –12

The following reaction takes place on dissolving silver cyanide in potassium cyanide:

$$AgCN + KCN = KAg(CN)_2$$

One mole of potassium cyanide combines with one mole of silver cyanide and all cyanide added in excess is present as free cyanide.

Silver cyanide is generally used to prepare fresh baths because it is readily soluble in potassium cyanide and can be obtained as a high-purity chemical.

The potassium cyanide that is required to form the potassium argentocyanide complex, is known as "combined" cyanide. The metal is furnished for deposition by ionization of this complex. The free potassium cyanide, added in excess of that required to render the silver cyanide soluble, favors a finer-grained deposit, aids in metal distribution and renders the anodes soluble at higher anode-current densities.

Potassium carbonate and potassium cyanide increase the bath conductivity. As potassium carbonate gradually builds up in the bath, a formula is selected which tolerates a wide range of carbonate concentration. In fact, the carbonate is beneficial and is always added to a fresh bath to increase the plating range.

Other chemicals, such as potassium hydroxide or potassium nitrate may be added to the cyanide bath. If potassium nitrate is added, a lower silver concentration may be used. In addition, the bath will tolerate a certain amount of sodium salts in place of potassium salts and some expense may be saved by use of a limited amount of the cheaper sodium salt.

Brighteners are often added to the bath to produce a finer-grained deposit and to reduce the tendency of the bath to plate trees on sharp-cornered articles. The brighteners commonly used are carbon disulfide and ammonium thiosulfate. If carbon disulfide is used, it cannot be added directly to the bath. The carbon disulfide must be shaken with a portion of the bath or a potassium cyanide solution. After standing for a day, a portion of this treated solution can be used as a brightener.

Ammonium thiosulfate may be added directly to the bath. Additions of brightener have to be made frequently and the quantity and effectiveness have to be determined for each bath formula.

For heavy deposits of silver at high current density, the following bath has been used:[2]

	g./l.
Silver Cyanide	45–50
Total Potassium Cyanide	45–50
Potassium Hydroxide	10–14
Potassium Carbonate	45–80
Potassium Nitrate	40–60
Temperature	102 to 113°F.

This bath contains both potassium hydroxide and potassium nitrate and is operated with ammonium thiosulfate as a brightener. The addition agent is added at 4-hour intervals with a total addition of 0.015 gram per liter every 24 hours. It is not necessary that the deposits are bright for engineering applications, but it has been found that the bright deposits have a far greater tendency to be free from pores. The conventional silver baths are operated at current densities of 5 to 15 amperes per square foot. The above bath may be operated at current densities up to 75 amperes per square foot if very rapid movement of the work or the solution is employed.

Preparation of the Bath

In preparing a silver bath, the potassium cyanide is dissolved first. The silver cyanide is then added to the potassium cyanide solution and the bath is agitated to promote solution. The other chemicals, which are all soluble, may then be added. The bath will probably contain some suspended material and some precipitate, unless very pure chemicals and distilled water are used. Therefore, the bath must be filtered prior to use. Immediate filtration is not always effective since the precipitates from hard water are colloid when first formed. However, these will coagulate and settle with time and can then be removed easily by filtration. For heavy, pore-free deposits, it is desirable to use continuous filtration.

Steel tanks may be used, but plastic-lined tanks are preferred in order to prevent troubles due to stray currents.

High-purity silver anodes are essential. If the anodes contain impurities which are insoluble in the bath, these will collect on the surface of the anodes as a black film. This black film will loosen at times and, since it is anodically charged, it will be attracted to the cathode where it will cause a rough deposit by occlusion in the plate. The anodes may be bagged to overcome this trouble, but operation is more difficult with bagged anodes. The anodes may also turn black if an excessive anode-current density is employed. This action is due to anode polarization. As the anode-current density is increased, a high-purity silver anode will turn from gray to brown to black. If the current density is too high and the anode is black because of polarization, the anode will begin to evolve oxygen gas. At this point, the bath will begin to go out of balance since the anode efficiency will drop and oxidation of the cyanide will take place. Normally, the anode and cathode efficiencies are both 100 per cent.

Preparation of the Basis Metal

The basis metal is prepared for plating by the usual cleaning and pickling procedures, but a silver strike is used prior to silver plating to obtain bond. The strike is usually operated at 6 volts and the silver concentration is low so that considerable gassing takes place while striking. Many different strike formulas have been proposed and, in some cases, a second strike is used and is operated at a lower voltage and with a higher silver concentration than the first strike. The strike deposit must not be too thick or it will be rough and it must not be too thin or it will redissolve as the work is lifted from the tank. The following strike [2] was found satisfactory for a striking time of 1 minute at 6 volts:

	g./l.
Silver Cyanide	0.75– 1.5
Sodium Cyanide	35 –45
Sodium Carbonate	0 –10

Sodium cyanide is used to make up the strike since the drag-in of a small amount of sodium salt from the strike is allowable in the silver bath.

In order to obtain maximum adhesion of silver to steel, it has been agreed [2,3,4] that a nickel plate prior to striking is required.

The process steps, particularly for plating heavy silver on steel, would be as follows:

1. Degrease
2. Electroclean
3. Rinse
4. Pickle
5. Rinse
6. Nickel-plate

7. Rinse
8. Use one or two strikes
9. Silver-plate

Hart and Heussner[3] also suggest alkaline anodic cleaning after anodic pickling as a further precaution. The work is then dipped in acid as the normal pickling step.

Steel parts, e.g., bearings must go into the silver strike with contact of a cathode lead to the part before it is immersed in the strike or the silver bath, and it must not be rinsed after striking.

Brass or copper may be plated by the same process as steel, except that nitric acid or an oxidizing pickle may be required in place of the sulfuric or hydrochloric acid commonly used on steel.

References

1. N. E. Promisel and D. Wood, *Trans. Electrochem. Soc.*, **80**, 466 (1941).
2. R. A. Schaefer, *The Monthly Review American Electroplaters' Soc.*, **33**, 1176 (1946).
3. J. S. Hart and C. E. Heussner, *The Monthly Review American Electroplaters' Soc.*, **33**, 142 (1946).
4. F. C. Mesle, *The Monthly Review American Electroplaters' Soc.*, **33**, 937 (1946).

Chapter 14

Acid Tin Baths

Tin is a coating for metals to be used in connection with foods such as for tin cans, machine parts for handling foods, refrigerator coils, etc. It is also used for a coating on hardware and small metal parts to be protected from corrosion while they are in storage.

Electrotinning has advantages over hot dipping in that the thickness can be controlled and better uniformity can be obtained over parts of irregular shape.

Tin coatings are popular because the appearance of the metal is pleasing and the coatings retain their color while offering good corrosion resistance. Tin is cathodic to steel, therefore, if the deposit is used over steel, corrosion resistance is obtained by producing a deposit that is of sufficient thickness to be essentially free from pores. On copper-base alloys, tin is anodic, therefore, tin will protect copper in the same way as zinc protects steel.

Formulation

Acid tin may be deposited from acid solutions. Tin baths are commonly prepared with stannous sulfate and sulfuric acid because of the availability and low cost of these chemicals. The following chemical limits are ordinarily used for tin and acid: [1]

	g./l.	*oz./gal.*
Stannous Sulfate	30– 90	4–12
Sulfuric Acid	30–150	4–20

Besides stannous sulfate and sulfuric acid, addition agents are also required. Experimentation has been carried on very recently with addition agents and changes in recommendations are still to be made.

Some of the addition agents are cresylic acid,[2] glue,[1] cresol,[1] β-naphthol,[3] resorcinol,[3] and many others. Two, three or four addition agents may be used and, of course, control becomes more difficult, as more addition agents are employed.

Pine [1] recommends the following specific bath:

	g./l.
Stannous Sulfate	54
Sulfuric Acid	100
Cresol Sulfonic Acid	100
β-Naphthol	1
Gelatin	2

In addition to these baths, proprietory baths [4] are also available.

Addition agents are essential to the operation of the bath. If they are not used, a loose mass of needles will be obtained on passing current for a short time. In fact, the tendency of the deposit to tree is used as a means of addition-agent control. When the deposit shows signs of becoming rough or treed in high current density areas, this is an indication that an addition agent is required. The kind and amount of addition agent may be determined by plating tests.

The bath has a tendency to decompose. The stannous sulfate is gradually oxidized to stannic sulfate which has a low solubility. Consequently, the bath continuously forms a precipitate.

The addition-agent problem and bath decomposition are definite disadvantages, but the bath has points in its favor. The anode and cathode efficiencies are both 100 per cent and the bath can be operated at room temperature. High plating rates can be obtained and high current densities are made possible by the use of agitation.

Function of Ingredients

The stannous sulfate is a source of metal ions and the sulfuric acid increases the bath conductivity. The variation in concentration of these two chemicals can be very wide. If higher plating rates are desired, then a high tin concentration should be used. However, no higher tin concentration should be used than is required, since drag-out and bath decomposition will be increased by an increase in tin content.

Bath Preparation

The order of addition of the chemicals is not important, but some of the addition agents are difficult to dissolve. Glue or gelatin should be treated with hot water to obtain a thick colloidal solution which may then be added to the bath. β-Naphthol has a low solubility, but it may be dissolved by heating in a portion of the bath. It may also be added to the bath as a concentrated alcoholic solution.[5]

The bath is highly corrosive so that plastic-lined or ceramic-lined tanks are required.

Bath Operation and Control

The bath has to be largely controlled by plating tests or by carefully watching the deposit for signs of need for addition agent. The amount of control is somewhat dependent on the thickness of the deposit required. In any plating bath, satisfactory flash deposits can be obtained with less control than that required to produce sound heavy deposits.

The major ingredients can be controlled by chemical analysis, but for many of the addition agents, analytical methods are not available.

For still plating, cathode-current densities of 10 to 40 amperes per square foot may be used. Anode-current densities should be kept below 25 amperes per square foot.

The sludge that forms in the bath does not present a problem since it readily settles. Occasional filtering may be required, however, if rough deposits are obtained. Filtering may remove a part of the colloidal addition agents, so that readjustment of the bath may be required after filtering.

Characteristics of the Bath

The acid tin bath has the typical characteristics of an acid bath. Throwing and covering power are poor, but this is offset by high electrode efficiencies which ease the operation.

Preparation of the Basis Metal

The preparation of the basis metal is similar to that for the acid zinc bath. Cleaning and etching are required and particular care must be taken that the metal is completely degreased since small spots of grease will not be covered by the deposit.

References

1. P. R. Pine, *Trans. Electrochem. Soc.,* **80,** 636 (1941).
2. U. S. Patent 1,397,222.
3. *J. Electrodepositors' Tech. Soc.,* **12,** 113 (1937).
4. Sulfate bath—Harshaw Chemical Co.; halogen tin bath—E. I. duPont de Nemours & Co.
5. F. C. Mathers, *Plating and Finishing Guidebook,* Metal Industry Publishing Co., New York, 1946, p. 115.

Chapter 15

Alkaline Tin Baths

Deposits from alkaline tin baths are used for the same purpose as those from acid tin baths, but the alkaline tin bath is the opposite, in many respects, of the acid tin bath. The bath must be operated hot, it has a very high throwing power, cleaning prior to plating is not so critical, and no addition agents are required.

Formulation

The following bath is recommended:[1]

	g./l.	*oz./gal.*
Sodium Stannate	120	16
Sodium Hydroxide	7.5	1
Sodium Acetate	15	2
Current Density	6–60 amp./sq.ft.	
Temperature	140–176°F.	
Anode Current Density	5–25 amp./sq.ft.	

The limits of the alkaline tin bath are almost self-defined. The plating rate is low compared to that of the acid bath. The plating rate can be increased by increasing the total tin concentration, but the solubility of sodium stannate is limited with build-up of carbonate in the bath. The following facts define the formulation and operation of the bath:

1. The cathode efficiency drops very rapidly with decrease in temperature.

2. The cathode efficiency drops with increase in caustic content.

3. The cathode efficiency drops with increase in current density.

4. Bath decomposition is increased with increase in temperature.

5. Bath stability is increased with increase in caustic content.

6. If sodium stannite is formed in the bath, it will decompose very rapidly and very rough or spongy deposits will result.

7. Sodium stannite is formed if the anodes are not properly polarized (white).

8. The anodes must be properly polarized (greenish-yellow to light-brown or gold in color) to obtain satisfactory bath performance.

9. If the anodes are excessively polarized (black), they will not be soluble.

10. The anodes will depolarize at a low anode-current density.

11. The anodes will become excessively polarized at a high anode-current density.

12. Continuous anode polarization is essential.

13. The anodes polarize more readily with increase in caustic content.

There are certain advantages that can be gained by changing any of the bath variables in either direction, but with each change, disadvantages also occur. The problem is really one of the magnitude of these changes and the operation must be governed by the known behavior of the bath. First, the bath must be operated at an economical plating rate and in order to operate continuously, the anode area must be adjusted to obtain the proper anode-current density.

The sodium stannate is a source of metal and the sodium hydroxide limits are self-defined. The sodium acetate acts as a buffer and aids in anode-polarization control. Sodium carbonate builds up in the bath with use and has only minor effects on the operation characteristics.

Bath Preparation

The bath is made up in a steel tank provided with heating coils.

The chemicals are soluble in water and they may be added in any order. Sodium stannate should be added in small amounts, because it is heavy and will lie on the bottom of the tank; and large amounts on the bottom of the tank are difficult to get into solution.

Operation and Control

After the bath is made up, it is heated to operating temperature. A good deposit should be obtained immediately and it is often well to start operation with steel anodes. The bath will operate well with steel anodes and after it is in operation, the tin anodes may be inserted one by one. By examination of the anode color, it will be obvious if the anode area of tin anodes should be increased or decreased. Some steel anodes may be used together with many tin anodes because the presence of steel anodes will prevent excessive polarization of the tin anodes. However, unless enough tin anodes are used at the proper anode-current density, tin will not be supplied to the bath at a sufficient rate. Steel anodes cannot be used exclusively and continuously because caustic soda will build up in the bath as tin is deposited. However, steel anodes are far less troublesome (and more satisfactory) for occasional plating and intermittent operation.

The anode problem and the cathode problem should be considered in view of the total current. For a desired plating rate, a certain current density will be required. The cathode area will then define the total current. From this total current and from the required anode-current density, the number of anodes can be calculated. As much tin as possible should be taken into solution since tin is cheaper in the form of anodes than as sodium stannate.

Analyses for tin and sodium hydroxide should be made several times a week, and additions of chemicals should be made daily on the basis of chemical consumption. If this is done, the fluctuation in caustic content will be small and anode control will not be difficult for continuous operation.

If sodium stannite is produced by depolarized anodes, the bath will turn dark and a rough or spongy deposit will be obtained. This is readily corrected by adding hydrogen peroxide to the bath. Temporary operation with steel anodes will be best during such a corrective period.

With steel anodes and hydrogen peroxide available, anode troubles are not serious. A knowledge of the bath behavior is all that is needed to operate the bath and as experience is gained, operation will become easier, since the experienced plater can tell what is happening in the bath from the number and color of the anodes and from the bath voltage.

Characteristics of the Bath

The major disadvantage of the alkaline tin bath is that high plating rates cannot be obtained. However, for most applications; a thin deposit is all that is required and a high plating rate is not essential. The problem of close anode control is a second disadvantage, but the alternative of operating an acid tin bath is not easy.

The major advantage of an alkaline tin bath is that covering and throwing power are excellent. The characteristic drop in cathode efficiency with increase in current density prevents build-up in high-current-density areas and at the same time promotes covering in the recesses.

Potassium Stannate Bath [2]

Tin may be deposited from an alkaline bath made from potassium stannate and potassium hydroxide. The use of potassium salts results in higher cathode efficiencies at higher current densities so that a higher plating rate can be obtained than with a corresponding sodium bath. Further, the potassium salts are more soluble than the sodium salts, they have a higher conductivity, and potassium stannate does not decompose as readily as sodium stannate. The cost of potassium salts is higher than that of sodium salts, but an increase in plating speed often means a decrease in plating cost.

Preparation of the Basis Metal

The basis metal is prepared for plating by cleaning and pickling. Cleaning is not as critical as for an acid bath since the hot alkaline tin bath has fair cleaning power.

References

1. F. F. Oplinger, *Plating and Finishing Guidebook,* Metal Industry Publishing Co., New York, 1946, p. 121.
2. F. A. Lowenheim, *Trans. Electrochem. Soc.,* **84**, 195 (1943).

CHAPTER 16

Acid Zinc Baths

Zinc is popular as a coating for steel because it offers good corrosion resistance at a low coating cost. Since zinc is anodic to steel, it will protect the steel from corrosion even though the deposit is porous or contains small breaks.

Zinc may be applied to steel by galvanizing, sherardizing or metal spraying, but electroplating is preferable for certain requirements. Zinc coating by electroplating, sometimes called electrogalvanizing, has the advantage that the thickness of the deposit can be easily controlled. Also, the electrogalvanized coating is free from brittle iron-zinc compound layers that are formed in a hot process. Because the coating is not brittle, it may be applied to a sheet that has to withstand subsequent forming operations.

Acid zinc baths are used where it is desirable to have a high plating rate and low cost. The deposits are not as attractive as those from the bright cyanide baths and the throwing power of the acid bath does not compare with that of the cyanide bath.

The acid baths are primarily used for coating wire and steel strip.

Formulation

A typical formula for an acid zinc bath is as follows: [1]

	g./l.	*oz./gal.*
$ZnSO_4.7H_2O$	360	48
NH_4Cl	30	4
$NaC_2H_3O_2.3H_2O$	15	2
Glucose	120	16
Current Density	10–30 amp./sq.ft.	
Temperature	75–85°F.	
pH	3.5–4.5	

Zinc sulfate is used as a source of metal; ammonium chloride increases the conductivity; sodium acetate acts as a buffer and glucose acts as an addition agent. Various formulations are possible and are equally effective as the one given. Other chlorides may be added to the bath, aluminum salts may be used in place of the ammonium salts or in place of the acetate and also various addition agents may be used. The chemicals selected should supply metal content, act as a buffer, increase conductivity and act as an addition agent. The same requirements hold for other acid baths, such as nickel, but the plating range of a nickel bath is readily changed by changing the formula. In an acid zinc bath, the formula may be selected with more freedom, depending on the availability, cost and purity of chemicals. Zinc sulfate is a high-purity commercial chemical made from a cheap commercial acid: sulfuric acid. The other chemicals in the formula given are produced on a large scale and are widely used in the chemical industries.

Bath Preparation

All of the chemicals used are readily soluble so that the order of addition is of little importance. After the chemicals are dissolved, the pH is adjusted and any required treatment of the bath is judged by the appearance of the deposit on initial plating. If the plate is rough, pitted or off color, this may be due to suspended or soluble impurities. Suspended impurities can be removed by filtration or by allowing to settle. Soluble noble-metal impurities can be removed by low-current-density electrolysis, by immersion plating on the anodes or by treatment of the bath with zinc dust.

Operation and Control

Although the bath is normally operated at current densities of 10 to 30 amperes per square foot, much higher current densities may be used if the bath is agitated. The current density is only restricted by the degree of agitation. For plating on wire, under special conditions, a current density as high as 2,000 amperes per square foot may be used. If the bath is agitated during rapid plating, it is important that it is kept free of insoluble material, therefore, continuous filtration is necessary. However, in still plating operations, occasional filtering or occasional removal of sludge from the bottom of the tank is sufficient.

The bath temperature should be maintained near room temperature, so that if large currents are used, cooling becomes necessary.

The bath is not sensitive to changes in composition, therefore, chemical control is not a major problem. Control of pH is the most important factor. The pH gradually rises due to reaction of the acid in the bath with the anodes. When the pH rises to the higher limit, sulfuric acid is added to restore the lower pH limit. The acid that reacts with the zinc anodes forms zinc sulfate at a rate which approximately balances the zinc lost by drag-out.

Anodes should be used that are high in zinc content and uniform in grain size for even anode corrosion. Zinc is an active metal and most anode impurities will collect on the anode as a sludge. If these are excessive, they may cause uneven corrosion and they may contaminate the bath with suspended impurities. In the latter case, the anodes may be bagged.

Characteristics of the Bath

The anode and cathode efficiencies of the bath are very high and the anode and cathode polarizations are low. Because of these conditions, bath balance and bath control are easy, but these conditions also cause the throwing power of the bath to be very poor. Thus, the bath is limited to plating simple shapes or to the use of special racking or anode arrangement to obtain good metal distribution.

Preparation of the Basis Metal

Steel is prepared for zinc plating by cleaning and pickling. As in other baths, the cleaning procedure depends on the condition of the steel, but care should be taken that all grease is removed from the steel since the covering power of the acid zinc bath is poor and the bath has no cleaning power. The steel should then be pickled long enough to produce an etch but it should not be over-pickled. If difficulty is experienced in covering the steel, a strike in a cyanide zinc bath or in a cyanide copper bath may be helpful.

References

1. W. Blum and G. B. Hogaboom, *Principles of Electroplating and Electroforming,* McGraw-Hill Bk. Co., New York, 1930, p. 316.

Chapter 17

Zinc Cyanide Baths

Zinc cyanide baths are used to coat steel for protection from rusting. Thus, while they are used for the same purpose as the acid zinc baths, they have certain advantages over the acid zinc baths. First, the throwing power is good, so that irregular pieces can be easily covered. Second, bright deposits can be obtained, and bright deposits have sales appeal, even in applications where they are not required. Also, bright deposits do not stain as readily as dull deposits so that they remain attractive longer. However, it must be kept in mind that zinc is an active metal. This very property that is responsible for its good protection of steel is also a property that will cause the zinc to lose its original appearance much more rapidly than other electrodeposits.

Table 6 gives a direct comparison between the properties of the acid and cyanide baths:

Table 6

Comparison of Acid and Cyanide Zinc Baths

	Acid Bath	*Cyanide Bath*
Throwing Power	Very low	Very high
Appearance	Dull gray	Semibright to bright
Basis Metal	Can plate all ferrous metals	Cannot coat malleable and cast iron
Plating Speed	Rapid plating is possible	Limited plating rate
Cost	Low	Low, but higher than acid baths
Tanks	Acid-resisting	Steel
Structure	Coarse-grained	Fine-grained
Control	Simple	Complex

TABLE 6 (*Continued*)

	Acid Bath	*Cyanide Bath*
Formula	Available	Proprietary for bright baths
Electrode Efficiencies	High at all plating rates	High only under limited conditions
Electrode Polarization	Low	High
Preparation of Basis Metal	Clean and pickle but use more care than for cyanide baths	Clean and pickle

The Chemistry of the Zinc Cyanide Bath

A zinc cyanide bath is prepared from zinc cyanide or zinc oxide, sodium cyanide and sodium hydroxide. Zinc cyanide combines with sodium cyanide to form sodium zinc cyanide.

$$2NaCN + Zn(CN)_2 = Na_2Zn(CN)_4$$

However, according to Blum and Hogaboom,[1] most of the zinc combines with sodium hydroxide to form sodium zincate.

$$2NaOH + Zn(CN)_2 = Na_2ZnO_2 + 2HCN$$

The chemistry of the zinc cyanide bath, and many of the other cyanide baths, is not as simple as represented by these equations. There is much known about the cyanide complexes and there is still much that is desired. For those interested in the complexity of these reactions, it is well to consult a paper written by M. R. Thompson.[2] For the operation and control of a plating bath, it is only necessary to know the chemicals used and the limits in which they must be held. For control purposes, we must know what the analytical method measures. For instance, in a silver cyanide bath, free cyanide is measured and controlled. In a cyanide zinc bath, the total cyanide is measured so that the analysis tells nothing about how the cyanide is combined. As previously pointed out, it is not necessary to know this. Yet it is convenient to know that the sodium zincate is present in a larger quantity than the sodium zinc cyanide, because one will then keep in mind that sodium hydroxide functions in a capacity other than to increase conductivity and pH.

Formulation

The basic zinc cyanide bath is described by Blum and Hogaboom.[1]

Plain Cyanide Bath

	g./l.	*oz./gal.*
Zinc Cyanide	60	8
Sodium Cyanide	23	3
Sodium Hydroxide	53	7
Temperature	104–122°F.	
Current Density	10–20 amp./sq.ft.	
Cathode Efficiency	90–95%	

This plain zinc cyanide bath will produce good deposits for protection of steel from rusting. If the bath is modified, by increasing the total concentration and by increasing the temperature, the plating rate can be increased.

If a mercury salt is added to a zinc cyanide bath, an alloy deposit will be obtained that has a more pleasing appearance than a deposit from the plain zinc cyanide bath.[3]

Zinc Mercury Bath

	g./l.	*oz./gal.*
Zinc Cyanide	37.5	5
Sodium Cyanide	22.5	3
Sodium Hydroxide	30.0	4
Mercuric Oxide	0.25	0.03
Temperature	86–122°F.	
Current Density	40 amp./sq.ft.	
Anodes	0.1–1% mercury	

The bath is controlled in a manner similar to the control of the plain cyanide bath except that one more constituent has to be controlled: mercury. If the mercury becomes high in the deposit, spots will develop on aging. If the bath is allowed to stand idle, mercury will immersion-plate on the anodes and the bath will eventually become depleted of mercury.

The zinc-mercury bath has good throwing power and good covering power and the deposit protects steel in the same manner as a plain zinc deposit.

Bright zinc deposits can be obtained by increasing the total salt concentration, by keeping the bath free of impurities and by the use of addition agents.[4]

Bright Zinc Bath

	g./l.	*oz./gal.*
Zinc Cyanide	60– 82	8 –11
Sodium Cyanide	19– 64	2.5– 8.5
Sodium Hydroxide	75–112	10 –15
Temperature	82–100°F.	
Current Density	10–50 amp./sq.ft.	
Anodes	High-purity zinc	

Function of Bath Ingredients

Zinc cyanide is used as a source of metal in the cyanide baths. It is rendered soluble and modified chemically by sodium hydroxide and sodium cyanide. These chemicals, in addition to forming complex zinc compounds, contribute to the bath conductivity. The ratio of sodium zincate to sodium zinc cyanide affects the plating-bath characteristics. The ratio of these complexes depends on the metal content and the hydroxide and cyanide content. The characteristics of the bath are controlled by holding the sodium hydroxide content within limits and controlling the ratio of sodium cyanide to zinc metal. This ratio

$R = \frac{\text{total NaCN}}{\text{Zn}}$ should be held within narrow limits to control the bath characteristics.

The ratio is held within the range of 2.2 to 2.6 for plain cyanide baths and of 2.0 to 3.0 for bright baths.

Hull and Wernlund [5] have shown how the bright plating range is shifted by the change in the ratio (R) and also how the cathode efficiency drops with increase in R. This data are shown graphically in Figure 19.

It should be noted that the ratio is calculated from the total cyanide. This is the cyanide content obtained by analysis and includes the sodium cyanide plus the zinc cyanide calculated to the sodium cyanide equivalent. It is calculated from a fresh bath in the following manner:

Bath:

	g./l.
Zinc Cyanide	60
Sodium Cyanide	23

$$\text{Zinc metal} = \frac{60\ \text{Zn}}{\text{Zn(CN)}_2} = \frac{60 \times 65.4}{117.4} = 33.4$$

$$\text{Total cyanide as sodium cyanide} = 23 + 60\frac{2\text{NaCN}}{\text{Zn(CN)}_2}$$

$$= 23 + \frac{60 \times 98}{117.4} = 23 + 50.2 = 73.2$$

$$\text{R} = \frac{\text{NaCN}}{\text{Zn}} = \frac{73.2}{33.4} = 2.19$$

Carbonate will build up in zinc baths with use—due to absorption from the air and decomposition of chemicals. Usually though, it does not become excessive since the build-up is rather slow and is compensated by drag-out. If cooling means are available, excessive carbonate may be removed by crystallization. The temperature at which it will crystallize depends on the total salt concentration and can be determined by cooling a small portion of the bath in the laboratory.

Bath Preparation

The bath is best prepared by dissolving the sodium hydroxide and sodium cyanide first. The zinc cyanide may then be added since it will be soluble in this solution. Other chemicals that are required may be added and the bath can be made up to volume. A plating test should then be made to determine the plating

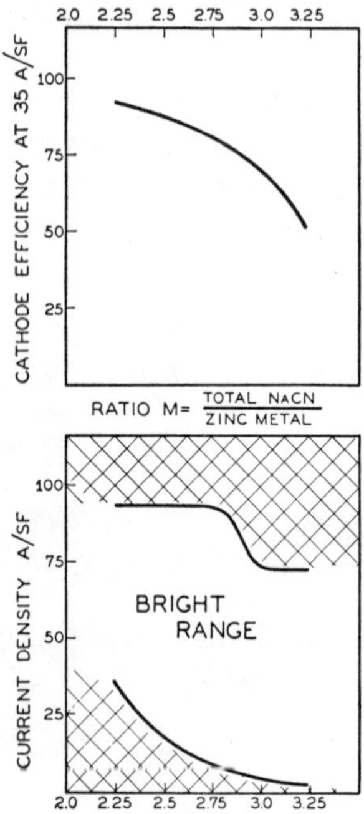

FIGURE 19

Variation of Current Density and Plating Range with Ratio

quality. The chemicals used for the zinc bath are of relatively high purity so that good deposits should be obtained immediately. If impurities are present, they will be revealed by an off-color deposit in the plating test. Or, they will be revealed by a shift in the plating range in the bright bath which is sensitive to impurities. Noble-metal impurities can be removed by electrolysis or more readily by treatment with zinc dust followed by filtration.

Operation and Control

The zinc cyanide bath is controlled by maintaining the sodium hydroxide content within limits and by adjusting the ratio of zinc metal to sodium cyanide. The difficulty of maintaining these limits will depend on the type of bath used, the application and the amount of work plated.

For higher plating rates and for barrel plating applications, higher chemical concentrations and higher temperatures may be used. However, the appearance of the plate will change with increase in temperature and if heavy currents are used, cooling will be necessary to keep the temperature from becoming too high.

The anode-current density should be held at 10 to 30 amperes per square foot to obtain proper corrosion. The anode-current density is obtained by adjusting the anode area, depending on the total tank current. Since zinc is chemically active in warm alkaline solutions, the anodes may go into solution faster than metal is removed at the cathode. This condition will be offset somewhat by the drag-out, but if it is excessive, it can be remedied by replacing some of the zinc anodes with steel anodes.

Bath balance depends on adjustment of the number and type of anodes so as to adjust the rate of solution of zinc to equal the rate of removal by plating plus drag-out. Sodium cyanide is then added to maintain the proper NaCN to Zn ratio.

Rapid plating conditions are the most difficult to maintain because the bath composition changes readily. However, such a bath can be maintained if frequent chemical and plating checks are made.

If the sodium cyanide content is high and zinc additions are required, zinc oxide may be added in place of zinc cyanide.

Preparation of the Basis Metal

The preparation of the basis metal is the same as for the acid zinc bath except that the cyanide bath has some detergent properties so that precleaning is not quite as critical.

References

1. W. Blum and G. B. Hogaboom, *Principles of Electroplating and Electroforming,* McGraw-Hill, New York, 1930.
2. M. R. Thompson, "Constitution and Properties of Cyanide Plating Baths," *Trans. Electrochem. Soc.,* 79, 417 (1941).
3. U. S. Patents 901,758; 1,435,875; 1,451,543; 1,497,265.
4. R. O. Hull and C. J. Wernlund, *Trans. Electrochem. Soc.,* 80, 416 (1941).
5. R. O. Hull and C. J. Wernlund, *Modern Electroplating,* New York, 1942, p. 368.

CHAPTER 18

Electroplated Alloys

In normal plating baths, only one metal is deposited and the electrodeposits are usually of high purity. However, if noble-metal impurities enter the plating bath, they will be codeposited with the major metal. Such impurities are a common source of trouble and result in rough, porous and off-color deposits. The behavior of these noble metals in a plating bath leads one to believe that it is not feasible to deposit more than one metal at the same time. In contradiction to this belief, there are cases where impurities are codeposited with the major metal without apparent interference. Small amounts of iron, for example, will codeposit with nickel in a Watts nickel bath without detection. Larger amounts will cause the deposit to be dull or burned. Appreciable amounts of cobalt may be codeposited with nickel. In fact, bright deposits are readily obtained, by codeposition of nickel-cobalt alloys in the presence of addition agents.[1,2] Iron-cobalt and nickel may all be deposited simultaneously in a variety of proportions.

Iron, nickel and cobalt serve well to illustrate the requirements for alloy plating. In order to codeposit metals, they must be compatible. That is, they must be compatible both from the metallurgical and the electrochemical standpoint.

Metallurgically, they must be metals that will form solid solutions or intermetallic compounds. The minor metal must fit into the lattice of the parent metal. This means that the most successfully plated alloys are the same ones that are successfully prepared by casting procedures. Examples of alloys successfully prepared by both procedures are: brass, bronze and solder. It must be kept in mind, however, that electroplating is a cold process as compared to casting. Therefore, the equilibrium at the plating temperature will be far from that at which alloys are formed from a melt. In order to predict the possible results to be obtained by electrodepositing an alloy, the phase diagram should be examined in the low-temperature region. However, predictions cannot be made with certainty for electrodeposits are often finely crystalline and fine crystal size may promote greater solubility than estimated from a phase diagram. The physical chemist knows that very small crystals are more soluble than large ones. When the surface area is large compared to the mass, deviations from the phase diagram may be expected. Lead, for example, has been codeposited with silver as a solid solution containing 10 per cent lead, which is much higher than expected according to the phase diagram. Codeposits form solid solutions and intermetallic compounds, but the structures will not be similar to cast structures. First, they will be more finely crystalline and second, certain structures will not be obtained at all. A eutectic structure cannot be obtained because it is formed from a melt and no parallel to such a freezing process takes place during deposition. Yet, in some cases, the structure of electrodeposits may approach the cast structures. If an alloy is cast, then cold-worked and annealed in order to obtain a fine structure and to attain low-temperature equilibrium, it may approach the structure obtained from an electrodeposit annealed at the same low temperature.

The points of comparison between cast and electrodeposited material are: the most successfully prepared alloys are those formed from metals that have an affinity for each other, but different results may be expected with different processes of formation.

The alloying metals must be compatible from the electrochemical standpoint.

Iron, cobalt and nickel are similar electrochemically. They all tend to deposit at approximately the same voltage. Therefore, codeposits of these metals are readily obtained. Since sound alloys can be prepared from these metals, an alloy deposit is obtained. An alloy deposit is, in this sense, defined as a sound deposit. Codeposits may be formed merely because ions of two metals are discharged at the cathode simultaneously, but the structure will be weak if the metals do not have an affinity for each other.

Iron, cobalt and nickel codeposit because they have a similar deposition potential. By referring to the electrochemical series, it will be seen that the equilibrium potentials of the metals cover a wide range. If the equilibrium potentials are far apart, the deposition potentials in solutions containing simple ions probably will not be favorable. Therefore, the number of alloys successfully deposited from acid solutions is considerably less than can be expected to be deposited from the complex-ion solutions. The potentials of lead and tin are close together so that they are expected to codeposit. Actually lead-tin alloys can be deposited from acid solutions. However, tin does not codeposit readily unless addition agents are used.

Alloy plating from solutions of simple ions is limited because of the deposition potentials of the metals. These potentials can be changed somewhat by changing concentration, acidity, temperature, current density, etc. But there is nothing that can be done to codeposit an active metal like zinc with copper from acid solutions. Small amounts of a nobler metal may be deposited, such as gold with copper, and copper with lead, but here again the codeposition is limited to small amounts. The reason for this is that the noble-metal plating rate depends on the rate at which it reaches the cathode. For instance, copper will codeposit with lead as fast as the copper diffuses to the cathode. If a lead cathode is used, the copper will continue to deposit even though the current is not flowing. In fact, copper will deposit on a lead anode while lead is going into solution under the influence of current. Thus the copper is not under the influence of the usual plating variables in the same manner as lead. If the acid bath contains large amounts of copper, it will not be possible to codeposit lead. We prefer to call a plating system of this type a "noble-metal system."

In any alloy system, one of the metals acts as a noble metal or, in other words, one metal tends to deposit more readily than the other. It is necessary then to use the term noble metal even in the potential system.

The noble-metal system is recognized by the fact that the plating rate of the noble metal is primarily dependent on the concentration of this metal in the bath. In other words, it is independent of the concentration ratio of the two metals.

The potential system is recognized by the fact that the percentage of metal obtained in the deposit is dependent on the ratio of the metals in the bath.

The noble metal system is limited, in solutions of simple ions, to small amounts of the noble metal in the deposit.

Most acid solutions provide simple ions and the deposition potential depends on the ionic concentration. If the ionic concentration can be changed then the tendency to plate (the potential) can also be changed, i.e., can be controlled. The potential can be changed about 0.06 volt by every 10 times change in concentration; so that if the potentials of two metals are wide apart it is not feasible to bring them together by change in concentration alone.

Ionic concentrations can be greatly changed by the formation of complex ions. These complex ions are most conveniently formed in cyanide solutions for plating purposes.

For instance, the following complex is formed in a silver cyanide bath:

$$KCN + AgCN = KAg(CN)_2$$

This complex ionizes as follows:

$$KAg(CN)_2 = K^+ + Ag(CN)^-{}_2$$

In this primary ionization no silver ions are formed. A secondary ionization takes place that gives rise to a small amount of silver ions.

$$Ag(CN)^-{}_2 = Ag^+ + 2CN^-$$

Cadmium, zinc, gold, copper, iron and a number of other metals form cyanide complexes. Tin and lead form complexes with caustic soda. If caustic soda is used to dissolve a tin salt, the tin-ion concentration will be so greatly reduced that tin will not deposit from a cold solution. Iron forms a cyanide complex that is so stable ionically that iron will not deposit on passing current across the bath.

By the formation of complexes with cyanide and caustic soda, the potentials of the metals can be greatly changed. If a copper salt and sodium cyanide are added to an aklaline tin bath, copper can be deposited. If the solution is operated cold, copper and hydrogen will appear at the cathode. The deposition potentials of copper and hydrogen then must be approximately the same under these conditions. If the solution is heated and a considerable excess of cyanide is added, then tin and hydrogen will appear at the cathode. Copper will not deposit even though a high concentration is present (with a great excess of free cyanide). By proper adjustment of conditions, tin and copper can be codeposited without depositing hydrogen. In fact, by proper adjustment of one, any two or all three ions can be deposited at the same time.

By the formation of complexes, it is possible to codeposit alloys that cannot be deposited from acid solutions. Tin and copper can be codeposited in all proportions from complex-ion solutions. Tin-copper-zinc alloys can also be deposited from these solutions.

Cyanides are the most satisfactory for formation of complexes and for successful alloy plating since many of the present plating baths are cyanide baths. By addition of caustic soda to the bath, further control of the deposition potentials is possible. Zinc and indium form complexes with both cyanide and caustic soda.

By the use of cyanide baths, copper can be codeposited with tin and zinc; silver can be codeposited with lead and cadmium. Deposition of many alloys is possible that are as yet unexplored, although hundreds of alloys have been codeposited.

Although many alloys have been deposited, only a few are used commercially. Brass is deposited quite widely. Solder, speculum metal and copper-tin-zinc alloy are deposited to a lesser extent.

There is interest in alloy plating, but it is not widely used in comparison with plating of the pure metals. The primary reason for this is that the development of alloy baths is difficult. There is more to the problem than merely bringing the potentials of the metals together. In order to be commercially successful, the bath must be stable, the salts used must have an appreciable solubility, the deposit obtained must be sound, and continuous control of the bath must be possible. The alloy must deposit at an appreciable rate, at the usual current densities, and at a sufficiently high cathode efficiency. Bath balance must be possible to maintain control. This problem includes the development of soluble anodes.

Wider use of alloy deposits is dependent on the demand. They are widely used and are successful as coatings for other metals. It remains to more fully demonstrate the value of alloy deposits. Among the advantages of alloy deposition, the following may be mentioned:

1. *Decorative Value.* Bright alloys can be deposited such as nickel-cobalt and copper-tin-zinc.

2. *Economy.* Certain alloys will be cheaper than pure metals used for the same purpose.

3. *Corrosion.* It is well known that the corrosion resistance of many alloys is greater than that of pure metals.

4. *Engineering.* Lead alloys are being successfully used as the running surface for bearings.

5. *Variable Properties.* Where an application is dependent on the properties of the coating, some control of the properties is possible by the application of alloys.

It is important that the metallurgist and the engineer consider alloy deposits where coatings are used. By realization of how the properties of alloys may be applied to coatings, the use of alloy plating will spread. No one knows better than the metallurgist how the mechanical properties of a coating may be changed by variation of the alloy composition. Wear resistance, hardness and ductility may be changed depending on the demand for greater wear resistance or better formability.

The brass bath is a good example of how a problem was solved by the use of an alloy deposit. In the rubber-tire industry, it was desirable to bond rubber to steel wires. After considerable experimentation, it was found that rubber could be bonded to brass and that thin coatings of brass could be bonded to steel by electroplating. Thus the rubber could be bonded to steel by the use of brass as a bonding medium. This application greatly stimulated the development of the brass-plating bath.

Electrochemistry of an Alloy Bath

An alloy-plating bath differs from a single-metal bath in that greater control is usually required. More than one metal is present in the bath and each metal will respond differently or the alloy will respond to a greater extent to changes in bath conditions. Since limits are placed on the percentage of metal in the deposit, the bath variables must be controlled to operate within these limits.

If we have two metals depositing at the same time, one will tend to deposit more readily than the other. This metal is called the noble metal. If there is an appreciable difference in the tendency of the two metals to deposit, the noble metal will deposit exclusively at low current density. As the current density is increased and as the diffusion layer at the cathode film thickens, it becomes more difficult for the noble metal to deposit in competition with the second metal present. Any variable that tends to increase polarization or to thicken the cathode film will decrease the relative ability of the noble metal to deposit. Polarization is decreased, the cathode film is made thinner and the noble metal deposits more readily as a result of the following changes:

1. Decrease in current density

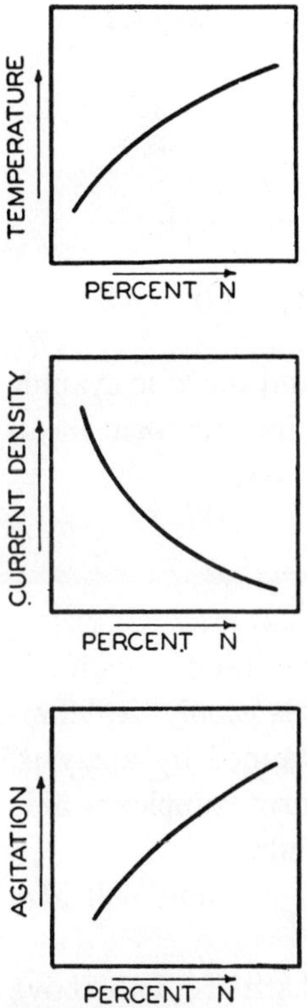

FIGURE 20
Effect of Variables on Alloy Deposits
N = noble metal in deposit

2. Increase in agitation
3. Increase in temperature

The variables are the same as those that fix the limiting current density in a single-metal bath.

If the limiting current density is low, it may be raised by raising the temperature or increasing the agitation. The extent of the change in noble-metal content of the deposit with the change in bath variables depends on the difference in tendency of the two metals to deposit. If this difference in tendency is great, then control may be difficult. If the difference is small the nobility may even reverse with change in conditions so that the second metal becomes more noble.

These rules are general and may not hold where temperature or chemical concentration affects the ionization of one metal more than that of another. By such variables, nobility may be reversed and the point of reversal is the ideal point or the point at which both metals will respond to the same extent to a change in current density.

Brass Plating

Brass is deposited to obtain adhesion of rubber to steel and for ornamental purposes. Various bath formulas have been proposed, but all commercial formulas are cyanide baths.

Formulation

The following bath has been recommended by Coats:[3]

	g./l.	*oz./gal.*
Copper Cyanide	26.2	3.5
Zinc Cyanide	11.3	1.5
Total Sodium Cyanide	45.0	6.0
Temperature	80–95°F.	
Current Density	9 amp./sq.ft.	
Brass Anodes	25% Zn	
pH (Colorimetric)	10.3–11.0	

The copper cyanide is a source of copper and the zinc cyanide a source of zinc. The sodium cyanide combines to form metal complexes of which the following are assumed:

$$2NaCN + CuCN = Na_2Cu(CN)_3$$
$$2NaCN + Zn(CN)_2 = Na_2Zn(CN)_4$$

All sodium cyanide, in excess of this combining amount, is defined as free sodium cyanide. However, this is only by definition since the free cyanide cannot be determined by analysis. The excess cyanide above that required to form complexes aids in anode corrosion and increases the bath conductivity.

Carbonate will build up in the bath with use and will also increase the bath conductivity.

The anode and cathode efficiencies of the bath are both above 75 per cent.

Bath Preparation

Steel tanks can be used satisfactorily for bath preparation and operation. Bath preparation consists of first dissolving the alkali cyanide and then adding the metal cyanides. The metal cyanides are soluble in and react with the sodium cyanide to form the complexes described previously.

Operation and Control

The bath must be operated to maintain the desired composition in the deposit. For decorative purposes, this may be done by color control as follows:

A series of brass alloys are prepared covering a range of

analyses, or a single brass plate of the desired color is used as a standard. The deposit is then compared with this color standard and if the colors do not match, the bath is corrected to obtain the proper color in the deposit.

For engineering purposes, the deposit is controlled by analysis.

Whether the deposit is controlled by analysis or color, a knowledge of the effect of the bath variables is essential.

If the deposit is high in copper, the percentage of copper may be decreased by:

1. Lowering the bath temperature
2. Adding caustic to raise the pH
3. Adding zinc cyanide
4. Adding both zinc cyanide and copper cyanide to lower the free cyanide content.

If the deposit is high in zinc the percentage of zinc may be decreased by:

1. Raising the bath temperature
2. Adding sodium bicarbonate to lower the pH
3. Adding copper cyanide
4. Adding alkali cyanide to increase the free cyanide content.

It is best to hold as many variables as possible within defined limits, but it is not possible to define all limits on an alloy bath. Since the limits on the deposit are defined, as many of the bath limits as possible should be set with at least one variable to obtain the proper analysis of the deposit. This is the main difference in operation of an alloy bath and a single-metal bath. In the single-metal bath, all limits can be defined. In cyanide baths, the cathode efficiency may then be a variable that will fluctuate depending on the condition of the bath at any given time. In the alloy bath, this condition cannot be tolerated. For instance, a drop in cathode efficiency in a brass bath may be due to a drop in the plating rate of the zinc alone. Consequently, the percentage of copper in the deposit will increase. Conversely, the plating rate of the copper may decrease and the percentage of zinc will increase in the deposit.

Another way of stating this same problem is as follows: The analysis of the deposit cannot be accurately predicted from an analysis of the bath. The cyanide bath changes continuously; both carbonate and ammonia continuously form in the bath due to cyanide decomposition. The ammonia is driven off in the alkaline bath, but it is not driven off at the same rate as it is formed.

The bath should be operated in such a manner that great changes or large additions do not have to be made to correct the composition of the deposit. This can be accomplished by making frequent checks, by making small corrective changes, and by defining the limits of total metal concentration. If corrections result in an increase of both copper and zinc, then the bath should be diluted, but great dilutions should be avoided. The problem is really one of bath balance, the same as it is for single-metal baths except that a balance has to be maintained for total metal and for the ratio of the metals in the deposit.

Temporary corrections can be made by changing the bath temperature or the current density. Increasing the current density will decrease the percentage of zinc in the deposit. This marks zinc as the noble metal in the recommended bath. However, in other brass baths,[4] an increase in current density has resulted in an increase in the zinc in the deposit. This means that the nobility of the copper and the zinc can be reversed, so that for a given bath, a log should be carefully kept and studied to determine the characteristics of the particular bath.

Plating-range tests will aid in control of the bath. If the deposit is off-color at a high current density, corrections may be made to obtain the proper color over a wide plating range. The deposit may turn red in the off-color range, a condition which may be due to either high copper or high zinc content. The metal responsible for the off-color deposit may be determined either by analysis of the deposit or by making an addition to the bath that favors deposition of one of the metals. If the test corrects the color, the cause of course, has been found. If the test does not correct the color, additions favoring deposition of the other metal should be tried. After the troubles have been located and corrected in the plating tests, the corrections are applied to the production bath.

Changes in the Bath

The cathode efficiency may be increased by increasing the bath temperature or by decreasing the free cyanide. These changes will also affect both the ratio of metals in the deposit and the anode corrosion, and should only be made in such a way that bath balance can be maintained.

Characteristics of the Bath

The cyanide brass bath has high covering and throwing power, but the deposit may go off color at extreme current densities. Therefore, closer attention to current distribution is required than for the single-metal cyanide baths if a uniform color is desired.

Preparation of the Basis Metal

The usual cleaning and pickling steps are required prior to brass plating.

Copper-tin Plating

Alloys of copper and tin are not widely used, but there is enough promise for such alloys that it is worth while to consider their deposition. We prefer to call a high-copper alloy from such a bath a bronze deposit, although high-copper alloys from the brass bath are sometimes also called bronze.

Copper-tin alloys may be deposited as bronze,[5] speculum metal [6] or tin-base alloys.[7] It is obvious then that high, intermediate or low copper can be deposited.

The behavior of the alloy baths of this type may be estimated from a knowledge of both the alkaline tin bath and the cyanide copper bath.

In the alkaline tin bath, the cathode efficiency is increased by increase in temperature, increase in metal content, and substitution of potassium salts for sodium salts. The cathode efficiency is decreased by an increase in caustic content.

In the cyanide copper bath, the cathode efficiency is increased by increase in temperature, increase in metal content and substitution of potassium salts for sodium salts. The cathode efficiency is decreased by an increase in free cyanide.

Since in each of these alkaline baths, a metal is discharged simultaneously with hydrogen, at about the same pH, it is obvious that the deposition potentials will be about the same.

If the two component baths are combined, the exact plating rate cannot be predicted because new chemicals have been introduced to each component bath. Also, the behavior of the anodes cannot be accurately predicted, nor can the rate of bath decomposition.

However, it can be predicted that tin will not deposit at low temperatures and that the tin plating rate can be decreased at high temperatures by increase in the caustic soda content; and the copper plating rate can be decreased by increase in cyanide concentration. Thus considerable control over the plating rate of each metal is possible by changing the bath composition or the plating conditions.

The final formula for a given percentage range in the deposit can only be determined by experiment to find the bath that can be operated continuously by the use of soluble anodes, without excessive bath decomposition.

Bronze

Bronze may be deposited from a bath of the following composition:

	g./l.	*oz./gal.*
Sodium Stannate	37½	7
Sodium Hydroxide	7½	1
Copper Cyanide	75	10
Total Sodium Cyanide	90	12
Temperature	170°F.	
Current Density	20 amp./sq.ft.	
Anodes	bronze	

The deposit will be of high copper content and the percentage can be changed by changing the bath composition or the plating conditions.

Bronze deposits are smooth and sound and can be plated heavily so that they may be used as a substitute for copper where such conditions are permissible or desired.

Speculum Metal

By decreasing the copper and total cyanide content and by increasing the tin content of the bronze bath, alloys containing 55 to 60 per cent copper can be obtained. This composition range approximates the compound Cu_3Sn and a very hard deposit is obtained. The deposit is bright so that it is attractive and resistant to wear. If it is deposited as a thin coating, it will stand some forming even though it is brittle.

The greatest possibility for a deposit of this type is for decorative purposes similar to the decorative applications of chromium.

Tin-base Alloys

There is little information available on the tin-base alloys, but it is known that tin-copper alloys can be deposited by adding copper cyanide and sodium cyanide to an alkaline tin bath.

The formulas that are available for plating copper and tin in various proportions are interesting because they indicate that the behavior of certain alloy baths can be estimated from a knowledge of the component baths. These formulas are also interesting because they indicate that widely different compositions can be deposited. Usually, when information on a plating bath is limited, it is not safe to expect too much from the bath. The widely used plating baths are those that have been developed through considerable experimental work and production experience. Nevertheless, it is not too difficult to make up a small bath and run preliminary tests to determine whether larger-scale operation is warranted. The alkaline cyanide baths containing copper and tin are certainly in a class worthy of exploitation.

Lead-Tin

In general, lead-tin deposits are used for the same purposes as lead, namely, the protection of steel against a corrosive environment. In addition, lead alloys are used as a running surface for bearings.

Lead-tin alloys may be deposited from the same baths which are used to deposit lead. All that is required is to add tin to the bath, to use lead-tin anodes and to control the solution composition to obtain the desired percentage of tin in the deposit.

Formulation

The lead-tin fluoborate bath is essentially the same as a lead fluoborate bath except that a second addition agent may be used along with gelatin.

The following bath was formulated by DuRose: [8]

	g./l.	*oz./gal.*
Lead	218.0	29.00
Tin	11.0	1.50
Glue	0.8	0.10
Resorcinol	0.4	0.05

A bath of this composition will deposit an alloy containing approximately 5 per cent tin. The bath is operated at room temperature with lead-tin anodes and at current densities of 10 to 40 amperes per square foot. The percentage of tin in the deposit will change with aging of the bath and also with the current density. Tin is not completely stable in any acid solution. It slowly oxidizes to stannic tin and since only the stannous tin is effective, a greater amount of soluble tin will be required to maintain the bath until an equilibrium is reached.

The operation and characteristics of a lead-tin bath are similar to those of a lead bath so that it is not necessary to go into the details again.

The bath need not be operated with a high lead concentration. High concentration will merely permit higher current densities. If low current densities are satisfactory, a more dilute bath may be used or if agitation is used, a more dilute bath may be used at high current density. If the bath is agitated, frequent filtration and bagging of the anodes may be necessary to avoid rough deposits. The concentration of chemicals and the operating current density are largely a matter of choice depending on thickness and quality of deposit and allowable plating time.

As little as 2 per cent tin in electroplated lead will give a more pleasing color than lead alone and will not darken as readily. The corrosion protection of steel is good, and DuRose claims that an alloy deposit containing 5 to 6 per cent tin will afford better corrosion protection of steel than pure lead or lead-tin alloys with greater percentages of tin.

Characteristics of the Bath

The anode and cathode efficiencies are essentially 100 per cent, which reduces the problem of bath balance to a minimum.

Some bath decomposition and sludge formation take place, but the amount is so small as to be insignificant.

The throwing and covering powers of the bath are poor, but are somewhat better than obtained in simple lead baths. The covering power may be further improved by the use of a copper strike since lead alloys will cover better over copper than they will over steel.

Equipment and Preparatory Steps

These are the same as for lead fluoborate.

Other Lead Alloys

Other lead alloys can be successfully deposited from acid baths as shown by the following fluoborate bath used for depositing lead-tin-copper alloy in bearing applications.[9]

	g./l.
Copper	1–3
Lead	100–110
Free Fluoboric Acid	35–45
Tin	11–13
Resorcinol	4–5
Gelatin	as needed
Temperature	room temperature
Current Density	20 amp./sq.ft.
Anodes	lead-tin

Bright Alloy Plating

The copper-tin-zinc alloy plating process developed by the Westinghouse Electric and Manufacturing Co.[10] is a good illustration of the advantages to be gained by alloy plating.

The plating solution is prepared, using copper cyanide, zinc cyanide, sodium stannate, sodium hydroxide, sodium cyanide and an addition agent, the total metal concentration being less than 1 ounce per gallon.

At a current density of 10 to 15 amperes per square foot the analysis of the deposit is approximately 56 per cent copper, 28 per cent tin and 16 per cent zinc.

The deposit is corrosion resistant and wear resistant, similarly to speculum metal, and in addition is bright.

Anodes of the same composition as the deposit are used. The anode efficiency is 100 per cent while the cathode efficiency is only 35 per cent. Therefore, a number of insoluble steel anodes must be used together with the soluble anodes to maintain bath balance.

The control of such a bath looks difficult when one considers that three metals are deposited at the same time from a bath of low metal content and at wide differences in anode and cathode efficiencies. Nevertheless, a number of tanks were operated and controlled successfully over a period of years.

A knowledge of the requirements of alkaline tin bath and cyanide copper and zinc baths is not as helpful here as in the case of bronze plating; although some of the principles of operation of the component baths will remain unchanged. A high pH must be maintained and the condition of the anodes must be watched as in an alkaline tin bath. The cyanide content must be frequently adjusted to control the copper-plating rate.

The bath is operated more as a strike bath than as most plating baths and it has the very high throwing power associated with strike baths. The low metal content is an advantage in that the bath is readily affected by solution of metal from the anodes. Since the anodes are of the composition desired in the deposit, they aid in bath control.

An 0.2 mil thick deposit gives very good protection against corrosion, and it is harder and more wear resistant than nickel. The deposit is nonmagnetic and is easily soldered.

Preparation of the Basis Metal

Preparation of the basis metal prior to plating includes cleaning and etching steps and, although the deposit may be applied directly over steel, it is recommended that copper is deposited first.

Nickel-Cobalt

Nickel-cobalt alloys are deposited for the same purposes as nickel, but they have the advantages of being harder and more corrosion resistant than nickel.

The following bath has been used widely for plating bright nickel-cobalt alloys:[11]

	g./l.	*oz./gal.*
$NiSO_4.6H_2O$	240	32
$NiCl_2.6H_2O$	45	6
Sodium Formate	35	4.5
H_3BO_3	30	4
$CoSO_4.7H_2O$	15	2
$(NH_4)_2SO_4$	2.5	0.3
Formaldehyde	1	0.12
pH	3.7	
Current Density	40 amp./sq.ft.	
Temperature	140°F.	

The nickel-cobalt bath is operated in a similar manner to a Watts nickel bath containing organic brighteners and the functions of the bath ingredients are about the same as those of a Watts nickel bath.

Organic addition agents are not necessary to obtain cobalt in the deposit; their function is to obtain a bright deposit. Sodium formate is said to act as a buffer at a pH of 3.7, while the boric acid is useful in increasing the limiting current density.[12]

Sodium formate and formaldehyde both act as brighteners, but agitation of the solution is essential to maintain deposit brightness.

The cobalt content of the solution, and thus of the alloy deposit, is controlled by the use of alloy anodes. These anodes maintain the nickel at its original concentration and, under steady operating conditions, will adjust the cobalt to the desired concentration.

The deposit is sensitive to suspended particles in the bath so that frequent filtration and bagging of the anodes is recommended. Since the bath is agitated, suspended matter will not settle to the bottom of the tank. Like bright nickel baths, the nickel-cobalt bath is sensitive to impurities and care should be taken to avoid contamination by copper, lead, iron, zinc and foreign organic matter. If impurities are encountered they must be removed by chemical means or by electrolysis at low current density.

Young and Struyk [1] investigated the effect of plating variables on the following nickel-cobalt bath:

	g./l.	*oz./gal.*
$NiCl_2.6H_2O$	100	
$CoCl_2.6H_2O$	5.5	
H_3BO_3	15–30	
Current Density	15–50 amp./sq.ft.	
pH	1.5–2.0	

The pH, current density, temperature and agitation effects were studied over a rather wide range. Over a large portion of these broad ranges, changes in conditions only had a minor effect on the composition of the alloy. The data obtained demonstrate the chemical and electrochemical similarity of nickel and cobalt. However, in certain critical ranges, such as at low pH, the plate composition was sensitive to small changes.

Smooth, ductile, fine-grained deposits were obtained from this bath without the use of addition agents.

Tin-Zinc

The tin-zinc alloys were studied by Angeles [13] and the deposits are claimed to be superior in corrosion protection to either tin or zinc alone. Tin is corrosion resistant, but if pores are present in the deposit, steel will rust at these points. Zinc, on the other hand, protects steel at the expense of self-corrosion so that the steel will not rust at pores in the deposit. Unfortunately, however, the corrosion rate of the zinc is not entirely satisfactory. By selecting the proper tin-zinc alloy, a deposit can be obtained that has the corrosion resistance of tin and yet steel will not rust at small pores in the deposit.

Experimental baths were made up containing 30 grams per liter of tin. These were operated at a current density of 15 amperes per square foot and at 70°C. The analyses of the deposits are shown in Table 7.

TABLE 7

Chemical Composition of Tin-Zinc Deposits

Grams per Liter			
Zinc	*Total NaCN*	*Free NaOH*	*% Sn*
1.0	15.0	2.5	92
1.8	22.5	3.6	86
2.5	24.5	4.3	78
3.6	26.0	4.8	70
8.0	41.0	5.0	53
12.0	53.0	12.5	28

Table 7 shows that a wide range of compositions can be obtained. For maximum corrosion resistance, a deposit containing 78 per cent tin was recommended.

The chemistry of the bath is complicated in that zinc forms complexes with both caustic soda and cyanide whereas the tin combines with the caustic alone.

The bath responds to the same variables that characterize an alkaline tin bath. The cathode efficiency increases with increase in tin content and temperature, and decreases with increase in current density.

Alloy anodes are used, but they must be maintained in a polarized condition. Anode efficiency decreases with increase in anode current density.

An increase in caustic soda content decreases the cathode efficiency and increases the zinc content in the deposit, showing that the caustic soda concentration controls the tin-plating rate more than it controls the zinc-plating rate. Changes in the sodium cyanide content, above a concentration of 20 grams per liter, has very little effect on the composition of the deposit.

The change in the composition of the deposit with change in current density illustrates the complexity of the bath composition. The percentage of zinc is a maximum at a current density of about 20 amperes per square foot.

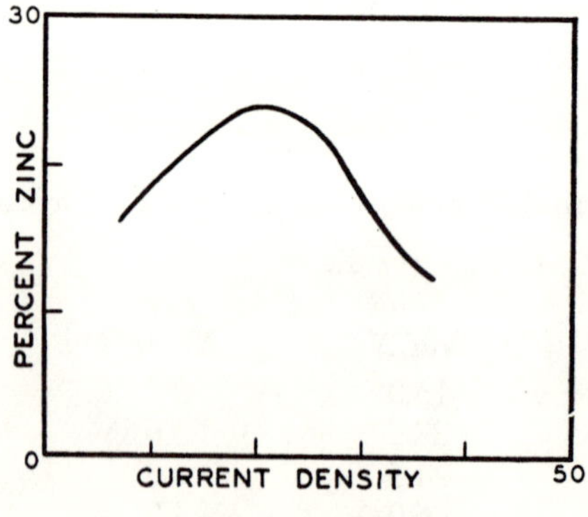

FIGURE 21

Current Density vs. Zinc Content of Deposit

The following bath is recommended to deposit 78 per cent tin:

	g./l.
Tin	30
Zinc	2.5
Sodium Cyanate	25
Sodium Hydroxide	4–6
Temperature	154–160°F.
Current Density	10–40 amp./sq.ft.
Anode-Current Density	7.5–15 amp./sq.ft.

References

1. C. B. F. Young and Clifford Struyk, *Trans. Electrochem. Soc.,* **89,** 383 (1946).
2. L. Weisberg, *Trans. Electrochem. Soc.,* **73,** 435 (1938).
3. H. P. Coats, *Trans. Electrochem. Soc.,* **80,** 445 (1941).
4. Ferguson and Sturdevant, *Trans. Electrochem. Soc.,* **38,** 167 (1920).
5. U. S. Patent 1,970,549.
6. International Tin Research Development Council, *Fourth General Report,* Pub. 95 (1939).
7. U. S. Patent 2,085,543.
8. A. H. DuRose, *Trans. Electrochem. Soc.,* **89,** 417 (1946).
9. R. A. Schaefer, *The Iron Age,* **159,** No. 15, 60 (1947).
10. M. B. Diggin and G. W. Jernstedt, *Convention Proceedings Am. Electroplaters' Soc.* (1944).
11. U. S. Patent 1,972,693.
12. L. Weisburg, *Trans. Electrochem. Soc.,* **73,** 435 (1938).
13. R. M. Angles, *J. Electrodepositors' Tech. Soc.,* **21,** 45 (1946).

Chapter 19

Copper Plating for Selective Carburization

The specifications for certain applications often call for certain areas of machine parts to be hard while other areas must remain relatively soft. The hard areas are produced by carburizing and the soft areas are allowed to remain in their original condition by protection from the environment during carburization.

A stop-off compound or a metal spray may be used for protection purposes, but copper plating is preferable as it produces a continuous protective coating.

Any metal that is impervious to carbon monoxide may be used to protect the desired area. Copper is used for such a protective metal because of its high melting point and because copper plating is inexpensive.

For a coating to be successful, it must be thick enough to exclude the carbon monoxide and also it must be free from breaks. If there are pores or cracks in the copper plate, hard spots will form during carburization that will interfere with machining operations or may even cause failure in service.[1]

A copper-plate thickness of less than 0.1 mil has been reported to protect steel, with a good surface finish, against a carburizing pack at temperatures of 1700 to 1740°F.[2] However, thicknesses of 0.5 to 5.0 mils are recommended.[3]

The thickness of copper required may vary with the type of copper bath employed. It has been reported that acid copper must be thicker than cyanide copper for the same degree of protection, but that there is little advantage in using bright copper.[4] The porosity of the deposit is dependent on the type of plating bath for a given steel and the same copper thickness. However, this is not as great a variable as the condition of the steel.

The condition of the steel is very important. It is well known that the porosity of a deposited metal depends greatly on the condition of the surface of the basis metal. If the steel is smooth, a thin coating may be pore-free; but if the steel is roughly machined or is a rough casting, a much thicker deposit will be required.

If the steel is rough, shot blasting may reduce the microscopic roughness and, consequently, reduce the thickness of the copper coating required for adequate protection.

A simple chemical test may be applied to the copper-plated part to measure the porosity of the deposit.[5] Such a chemical test is a rapid means of checking the effective coverage of the copper coating. It is good practice to examine the steel for faults before plating and then to examine sections of a finished part after carburizing. Examination of the steel may reveal gross defects such as occlusions in the surface, burrs from machining or cracks or pores in the metal. Any of these faults will be a source of porosity and thus of hard spots. Examination of a section after carburizing may reveal hard spots which can then be traced to a typical fault. Examination after carburizing may also reveal troubles in the plating practice such as failure to obtain bond because of imperfect cleaning, or incomplete removal of scale in the pickling step.

Cracks

If the steel is cracked or porous, troubles may arise during plating and hard spots may result from imperfect plating. The plated deposit may bridge small cracks, but if it does, a blister may develop at such a point on subsequent heating. If the crack is large, the plating may not bridge the crack, resulting in local failure.

Cracks may be found by microscopic examination of the

surface of the steel. If they are not found prior to plating, they may be revealed after plating through spots on the deposit.

Poor Bond

If the cleaning or pickling operations are inadequate, bond of the deposit to the basis metal may not be obtained. Spotty bond will be indicated by blisters on heating and total lack of bond will be revealed by loosening of the entire deposit.

Scale

If the original steel contains surface scale and this is not removed by pickling, then the deposit will not cover at these points and areas for development of hard spots will be present. Any nonconducting inclusion, such as slag or sand, will react in a similar manner.

Slivers or Burrs

Gross irregularities in the steel, such as slivers from a grinding operation or burrs, will not cover completely on copper plating and consequently, will result in an imperfect product.

Examination of Sections

Microscopic examination of a section after carburizing may reveal the source of the many troubles that lead to hard spots. If the deposit is blistered, sections should be taken through the blister. A crack is occasionally found beneath the blister, but if the blister is large, it is more likely that the steel was not clean prior to plating.

Hard spots may be found at a break in the coating because of small particles of nonconducting material. On close examination, it can be usually determined if this is scale, slag or sand.

At other hard spots, slivers of steel from poor grinding or burrs may be detected.

Figure 22 diagrammatically indicates the appearance of sections from various faults as found by examination after carburizing.

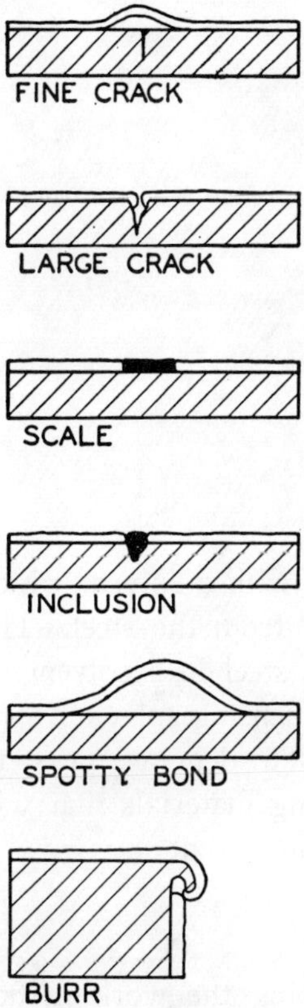

FIGURE 22
Causes of Poor Carburizing Results

Faults revealed by examination of sections after carburization will suggest steps to avoid further trouble. If the trouble is due to faulty steel, there is little that can be done except to obtain good steel at the start. If the steel is good but excessively rough, mechanical finishing steps will reduce the amount of copper plate required. However, machining operations are expensive so that it is usually more economical to deposit a heavier copper plate to compensate for the rough surface. Extreme irregularities such as burrs, slivers or undercut machining marks must be removed by shot blasting, tumbling or some other means.

Masking

In order to confine the copper plate to the desired areas, some method of masking the steel is required. If the areas to be protected are flat, the pieces may be stacked and a number plated at a time. This stacking procedure can be applied very seldom, because the areas to be protected do not often mate on stacking.

For large production runs, it is convenient and rapid to have rubber masks made that can be snapped on. For smaller runs, the pieces can be masked with masking tape, wax [6] or lacquer.

Plating Practice

The following steps constitute good plating practice:

1. Degrease
2. Mask
3. Alkali clean
4. Rinse
5. Pickle
6. Rinse
7. Cyanide copper plate
8. Rinse
9. Acid copper plate
10. Rinse
11. Dry

Degreasing

Degreasing operations are carried out with a solvent which will remove most of the grease and oil from the steel. The degreasing may be done by dipping the steel in a solvent or better still by degreasing the steel in condensing solvent vapor in a degreasing machine. It is necessary to degrease prior to masking since there are very few masking materials thai will not partially dissolve or be blistered in the solvent.

Alkali Cleaning

Alkali cleaning may be done by soaking the work in hot alkali and soap solution or by an electrolytic process. Hot electrolytic alkali cleaning is usually preferred as it is faster than soak cleaning. Electrolytic cleaning may be done with the work being either anode or cathode. Anodic cleaning is often preferred since metal impurities that may build up in the cleaner will not deposit on the work under these conditions.

Pickle

In most pickling operations, either hydrochloric or sulfuric acid is used at various temperatures and concentrations. Pickling times should be long enough to remove all rust and scale and to etch the surface of the steel. Anodic pickling may be used to hasten etching, but it will have little advantage over straight pickling for the removal of rust and scale.

The work must be handled rapidly after pickling to prevent rusting of the active steel with resulting poor adhesion of the copper deposit.

Cyanide Copper Plate

If a copper deposit up to 2 mils in thickness is sufficient for masking purposes, such a thickness may be satisfactorily obtained in a cyanide copper bath. For thicknesses greater than 2 mils, the deposit from the cyanide bath becomes too rough to be of value.

If a greater thickness is required, a thin cyanide copper plate of 0.01 to 0.05 mils is first applied. Such thicknesses are sufficient to insure bonding of a subsequent acid copper deposit.

Acid Copper Plate

Acid copper is deposited over cyanide copper to obtain heavy copper deposits. Smooth, sound deposits of almost any thickness can be deposited from the acid bath. The acid copper, however, cannot be applied directly to the steel since it will not bond.

References

1. A. D. Sayers, *The Monthly Review, American Electroplaters' Soc.*, **33**, 1271 (1946).

2. J. C. McCullough and O. M. Reiff, *Ind. Eng. Chem.*, **16**, 611 (1924).

3. R. M. Burns and A. E. Schuh, *Protective Coatings for Metals,* Reinhold Publishing Corp., New York, 1939.
4. M. M. Thompson, *Metal Finishing,* **40,** 579 (1942).
5. A. Brenner, *Plating and Finishing Guidebook,* Metal Industry Publishing Co., New York, 1945, p. 169.
6. C. E. Ernst, *Metal Finishing,* **44,** 155 (1946).

CHAPTER 20

Diffusion Coatings

Many methods of producing diffusion coatings have been developed.[1] For example, steel is treated with aluminum, carbon, chromium, nitrogen, silicon, tin and zinc by calorizing, carburizing, chromizing, nitriding, ihrigizing (siliconizing), stannizing and sherardizing. The various treating methods may be carried out with powdered metal, molten metal, a salt or a gas, each method producing a surface with properties different from those of the basis metal. The object of these treatments is usually to produce a more wear-resistant or a more corrosion-resistant surface. These thin coatings are applied then for the same purpose as electrodeposits are sometimes applied. The properties of these diffusion coatings depend on the extent and type of alloy formation. Such alloys can be formed by electroplating a metal on a basis metal and then subjecting the work to a heat treatment. These methods are occasionally used, although they are not widespread.

Corrosion-resistant coatings are formed by diffusion of nickel and zinc deposits applied over steel.[2] The coatings are claimed to be more corrosion-resistant than either of the single metals. This process, known as Corronizing may also be applied to nickel-tin. Corrosion-resistant coatings for bearing applications have been produced by diffusion of lead-indium; indium electrodeposits have also been diffused into cadmium, silver and copper.[3] In addition, many diffusion studies have been made of the alloys formed by heat-treatment of various electroplates over other metals.

The number of possible alloys that can be formed is great. Table 8 shows expected alloy formation from a study of a number of binary phase diagrams.

TABLE 8

Possible Alloy Formation from Binary Phase Diagrams

	Pb	*Ag*	*Cd*	*Cu*	*Sn*	*Al*
Fe	1	1	1	2	3	3
Ni	1	1	2	2	2–3	3
Zn	1	2–3	2	2	1	2
Pb	2	2	2	2	2	1
Sn	2	2–3	2	2–3	2	2
Ag	2	2	2–3	2	2–3	2–3
Cu	2	2	3	2	2–3	2–3

1 = No appreciable alloy formation
2 = Solid solution
3 = Compound formation

Solid-solubility or compound formation must be possible for diffusion to take place. When diffusion does occur, the phases formed will be as indicated in the low-temperature region of the phase diagram and they will appear in the same order. All the single phases in a binary diagram will appear as layers but the two phase regions will only appear as lines separating the single phases.

The following diagrammatic cases will illustrate the relationship between a phase diagram and a diffusion alloy.

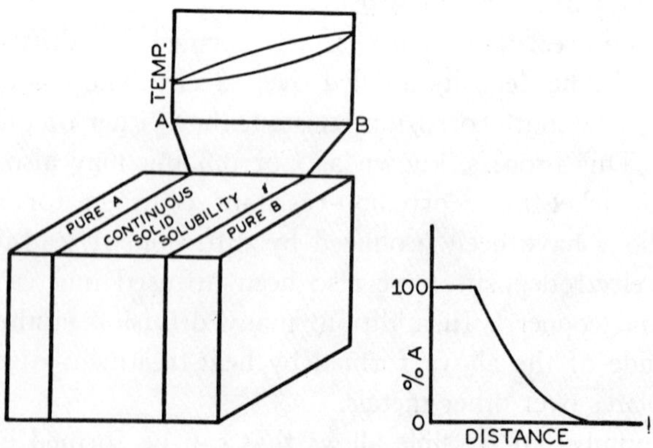

FIGURE 23

Continuous Solid-Solubility

Solid-Solubility

When two metals form a continuous series of solid solutions, one of the metals will diffuse into the other. Thus an alloy zone will be formed that will range from 0 to 100 per cent metal. If diffusion continues until the pure metals are used up, then the alloy zone will homogenize and form an alloy of a percentage depending upon the amount of each metal originally present.

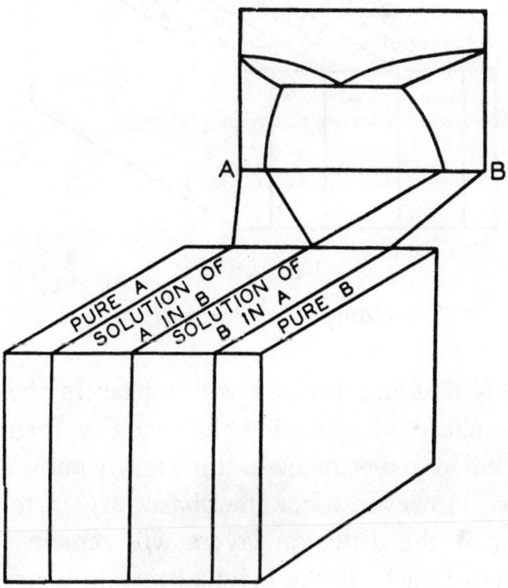

FIGURE 24
Limited Solid-Solubility

When two metals form solid solutions of limited solid-solubility, the method of diffusion will be similar to that for continuous solid-solubility except that two phases will be formed.

Compound Formation

If compounds or intermediate phases are indicated on the phase diagram, then these will also be formed on diffusion.

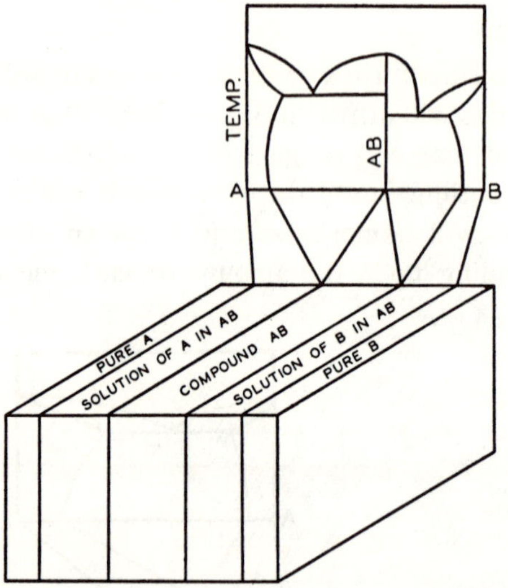

FIGURE 25

Compound Formation

The phases that are formed will appear in the order indicated on the phase diagram, but the quantity formed can only be determined by experiment or through a knowledge of diffusion rates. However, after the phases are formed, the ratio of thickness of the diffusion layers will remain the same as long as some of each of the original metals is present. After one of the metals is completely used up, homogenization will tend to take place. If homogenization does take place, then only one phase will remain, but this condition may take a very long time to occur. Small amounts of the phases are formed very readily, but as the thickness of the diffusion layers grows, any further increase in growth is slow. The thickness of the diffusion layer will be increased approximately as the square root of the time; so that, to double the thickness, the diffusion time will have to be increased four times.

Diffusion can take place at temperatures appreciably below the solidus on the phase diagram. Low-melting metals, like tin and indium, may even diffuse slowly into some metals, such as lead, at room temperature. Of course, the rate of diffusion under these conditions is slow, but it can be greatly accelerated by increasing the temperature. The diffusion rate will increase as the diffusion temperature approaches the melting point of the lower-melting of the two metals. If the solid-solubility of the two metals concerned is broad, then diffusion is likely to go on at a fair rate at elevated temperatures and a broad diffusion zone is possible. If three metals are diffused at the same time, then diffusion to form a ternary system may be more rapid than with two metals. This would be particularly true if the three metals were appreciably soluble in one another and would melt at a lower temperature than any of the three possible binary combinations.

Many new alloy combinations are possible by taking advantage of diffusion alloys produced from electrodeposits. The alloys that result may be homogenized or they may be in layers. If they are in layers, the properties will be different from those of alloys produced by casting or by electroplating.

Further study is required on diffusion in general and on diffusion of electrodeposited layers, but it is certain that many alloys, such as copper-zinc, copper-tin, copper-nickel, silver-zinc, silver-tin, lead-base, tin-base, zinc-base and many others, can be produced.

References

1. F. N. Rhines, *Surface Treatment of Metals,* American Society for Metals, Cleveland, 1941, p. 122.

2. G. Black, *Metal Finishing,* **44,** 207 (1946).

3. A. A. Smith, American Institute of Mining and Metallurgical Engineers, Institute of Metals Division, *Technical Publication No. 1640,* 1944.

CHAPTER 21

Selection of a Plating Bath

Electroplated coatings are widely applied for decorative and corrosion-resistance purposes. When used for such purposes, in volume production, the type and thickness of deposit is usually specified. Specification plating in volume production requires continuous testing to determine if standards are being met. Technically the problem is simple for all that is required is to manufacture a product that will meet standard tests. Tests for thickness of deposit on significant surfaces and tests for adhesion of deposits are not simple, but a number of acceptable routine tests are available. The manufacturer and the customer usually decide on appropriate tests for these and other properties, after which the manufacturer puts his platers, analysts and technicians to work to produce plating to specifications.

When the electroplating applications are of a miscellaneous nature, various problems are presented. For example, if it is required to plate a part for temporary protection against corrosion, considerable experimentation may be in order. First of all, decisions must be made regarding the probable exposure life of the part. The exposure conditions may be varied, for instance, the parts may be replacement parts for a mechanical device and they may lie on the distributor's or salesman's shelf for several years. The customer usually is familiar with the conditions to which the parts will be exposed and can satisfactorily set up the requirements for the product.

Another factor to be considered is the importance of the appearance of the electroplated part. Attractively finished parts have greater sales appeal even if the parts go in the interior of an appliance where they are very seldom seen. The economical aspects of plating the part is also an important factor.

In addition to corrosion resistance, appearance and economy, the plating process has to be compatible with the rest of the operations going into the manufacture of the part. If, for example, part of the metal is machined away during manufacture and a part of the deposit is returned for remelting with the metal scrap, the deposited metal must not be objectionable in such a manner as to contaminate the alloy. If the deposited metal is detrimental to scrap recovery it will be necessary to develop a means of removal of the metal prior to reclaiming the scrap material.

The plating process itself must also be considered. Is the process adaptable to the part? What will the cleaning and pickling steps do to the part? Do recessed areas have to be covered? Is selective plating called for? Can the deposit be bonded to the metal or alloy in question? These questions and many others can only be answered by complete understanding of the problem under consideration.

Plating can be used to a greater advantage through a knowledge of the properties of the individual metals and a knowledge of the characteristics of the plating baths. For instance, chromium is widely used to build up worn machine parts. Salvage of tools and dies by such a process makes it possible to reclaim old tools for a fraction of the cost of new ones. But, contrarily to general opinion, heavy chromium deposits are not cheap to produce. In the operation of chromium plating baths, very careful attention must be given to racking of the work. In addition, the plating rate is limited by heavy requirements of current and also by the extremely low cathode efficiency.

Hard nickel and iron deposits will often serve for tool salvage just as well as chromium and their application should be given every consideration. These baths are easier to operate, the racking problems are not as difficult and rapid plating rates are more easily obtained. When heavy deposits are produced that require days of plating, the plating rates determine the floor space required in the plating room, which, in turn, is related to overhead.

Plating of thin deposits may be used for many purposes in a large plant. Machined parts often rust if they wait more than a few days for the next operation, or they may even rust in a matter of hours if the temperature and humidity are high. The application of a thin electrodeposit may protect the parts from rust and the savings realized in reduction of scrap will more than pay for the extra step involved.

Dip-coatings of certain organic substances are more economical than electroplated coatings for temporary protection, but it is possible that plating can be applied at such a phase of the process as to be of more permanent value. Also the plated coating may make subsequent machining or handling operations easier. A coating may be selected that will not retain chips as readily as an oiled surface, or one may be selected that is easily soldered in a subsequent manufacturing step. If necessary, the deposit may be removed by a stripping method later in the process of manufacture. It is conceivable that plating may even be used as a means of identification of special parts. In large-scale manufacturing, special parts are occasionally run and there is a great danger of getting them mixed with regular production. For example, if all regular production were plated with zinc and special production with copper, the problem of segregation of the two parts would be reduced to a minimum.

In a plant that has laboratory facilities, it is convenient to supply plating service to the plating department. The average laboratory is equipped with hood, beakers, chemicals and most of the equipment required for small-scale operations. If a 50-ampere, 9-volt generator, a few ammeters and rheostats are added to the existing laboratory equipment, an appreciable amount of experimental plating can be done. If the plant personnel is familiarized with the possibilities of plating, then problems will begin to reach the laboratory.

When a well-defined problem is given to the laboratory, it

is good practice to prepare several plating baths and run some plating-range tests. Even if the baths are well known, nothing reveals the characteristics better than simple plating tests at various temperatures, over a range of current densities. If the problem is of sufficient importance, the tests may be extended to determine the effect of variation in chemical concentration and pH of the baths. This procedure of thoroughly studying the plating baths has further advantage in that the original plating-range standards can be used for future bath control. By using the plating-range test, small baths can be prepared and maintained by test. If they have sufficient life, it is easier to discard them and prepare new baths than to control them by analytical methods. However, some baths cannot be used fresh. Nickel, for instance, usually requires some working or purification before use. The same steps will be required for small laboratory baths as for production baths, except that the small baths can be made, filtered and electrolyzed in a matter of hours, whereas it may take days to do the same on a large scale.

The next problem in experimental plating is to plate some actual production pieces as this brings in the cleaning, etching and testing steps. In every new plating problem, a test for adhesion is probably the number one step. For very thin deposits up to 0.05 mil, used for temporary protection, poor adhesion may be permissible, but in practically all other cases, it is necessary to obtain bond.

A good way to test for bond in the laboratory is to deposit 10 to 30 mils of metal and then attempt to remove the deposit by mechanical means. Bending, twisting and pounding tests are often used; and a hammer and chisel can be used to learn much about heavy deposits.

The last stage of investigation consists of some type of test to predict the probable behavior in service. These tests include corrosion tests, wear tests, hardness tests and they are used as a basis for recommendation. If the recommendation leads to plant trial, then a pilot plant can be set up to determine

TABLE 9

Properties of Ccmmon Plating Baths

Metal	*Bath*	*Plating Range*	*Plating Rate*	*Throwing Power*	*Hardness*	*Corrosion Resistance*	*Control*
Brass	Cyanide	Wide	Slow	Good	Fair	Good	Careful
Bronze	Cyanide	Fair	Fair	Good	Fair to high	Good	Careful
Cadmium	Cyanide	Wide	Slow	Good	Low	Poor	Average
Chromium	Acid	Narrow	Fair	Poor	High	Excellent	Easy
Copper	Acid	Narrow	High	Poor	Fair	Tarnishes	Easy
Copper	Cyanide	Wide	Low	Good	Fair	Tarnishes	Average
Copper	High efficiency cyanide	Wide	Fair	Good	Fair	Tarnishes	Average
Lead	Acid	Narrow	High	Poor	Low	Oxidizes	Careful
Nickel	Acid	Fair	High	Poor	Fair to high	Good	Average
Silver	Cyanide	Wide	High	Good	Fair	Fair	Easy
Tin	Acid	Narrow	High	Poor	Low	Good	Difficult
Tin	Alkaline	Wide	Low	Excellent	Low	Good	Easy
Zinc	Acid	Narrow	High	Poor	Low	Poor	Average
Zinc	Cyanide	Wide	Low	Good	Low	Poor	Average

adaptability of the plating process and service tests will determine the acceptability of the product.

If an unusual plating problem emerges, the possibility of baths other than the common plating baths should be considered. Under the proper conditions, other metals can be deposited. If nickel and iron are candidates in the problem, cobalt should be certainly considered. If a very hard surface is required, a tungsten alloy may be worthy of consideration. Arsenic, antimony and aluminum can be deposited for experimental work. Many alloys and diffusion coatings are possible. It should be kept in mind, however, that these processes are not usable on a large scale without considerable experimental work and also that an experimental bath may fail for what appears to be a minor point in the laboratory. An experimental bath may fail because the bath is not stable, the plating range is narrow, the anodes are not soluble or even that analytical methods are not available or cannot be developed. If an addition agent problem exists, hundreds of tests may be required.

Experimentation with new and unusual plating baths should only be undertaken if it is permissible to assume an extensive laboratory and pilot-plant investigation.

Most plating problems should be attacked by starting with the standard common plating baths. Some of the properties and characteristics of these baths are listed in table 9 as a guide for selection.

CHAPTER 22

Continuous Plating

Maximum economy is reached in plating operations by the use of continuous plating machines. These machines are usually referred to as automatics and the plating operations may be of a semiautomatic or fully automatic nature.

Automatic Plating Machines

Fully automatic operation is, of course, the ideal plating condition. In fully automatic plating, the racked unplated work is suspended from hangers at the beginning end of the machine and the finished plated product is removed at the other end. Before entering the machine, the work is usually dirty and unetched, but the operations are so complete that the product at the exit end is fully plated and dry, ready for packing. Other steps, such as bright-dipping and passivating may be added to the cycle, depending on the requirements of the process and the economy involved. A double rinse or a combination spray and running rinse are sometimes of value. If a high concentration of metal is used in the bath, a drag-out tank may be installed from which metal may be reclaimed, and also strike steps may be added. A flow sheet may read as follows:

1. Degrease
2. Electroclean
3. Rinse
4. Acid-dip
5. Rinse
6. Strike
7. Plate
8. Drag-out

9. Rinse
10. Dry.

Such a process would be typical for plating steel with a high-efficiency copper bath; or for plating steel- or copper-base alloys with silver.

In plating zinc-base die castings with copper, nickel and chromium, a semiautomatic machine may be used, since it is often necessary to buff the pieces either after copper plating or after nickel plating. Thus, the process has to be divided into two parts, each using an automatic machine. One automatic would deliver preplated pieces to the buffing department which, in turn, would send them to a second automatic plating machine for finish-plating and inspection.

When plating is put into automatic operation, improvements in the process and improvements in control become attractive. A log on a continuously operating fully automatic machine will tell just what is happening in the process. The quantity of metal consumed per day can be closely observed and the metal going to drag-out and recovery can be easily checked on a large scale. With accumulation of data, improvements can be made in rinsing and drying operations and etching may be changed from chemical to electrolytic etch for closer control.

Automatic plating is applied on an even larger scale where metal is plated as strip or wire before fabrication. The large number of patents that have been issued on various improvements of such machines indicates the detail in which such processes have been studied. Electrotinning is a good example of such a large-scale process.

Electrotinning

Electrotinning replaced hot-dip tinning in many plants a few years ago because of a demand for thinner tin deposits. This was essential on account of a tin shortage and because the hot-dip process was limited to a minimum thickness of 1¼ pounds per base box (0.075 mil). The electrotinning process has the advantages of producing, at high rates of speed, uniformly distributed deposits of any desired thickness. A thickness of ½ pound of tin per base box was used on a great deal of tinned sheet for the canning industry. This thickness of tin (0.03 mil) does not afford the same protection as 1¼ pounds per base box obtained by the hot dip process, but it was made to serve the purpose by enameling the cans.[1] The enameling operation is expensive, therefore, it is assumed that heavier tin coatings will be used for protection against the corrosive action on the can of foods such as spinach, peas and tomato juice and that the thinner tin deposits will be used to fabricate containers for less corrosive conditions such as containers for coffee, oil and paint.

An electrotinning pilot plant was run as early as 1934 and full-scale production was introduced by 1936 by the Carnegie Illinois Steel Corporation.[2] This process, known as the "Ferrostan" process, used an acid tin bath. The plating line was 230 feet long and could be operated at a strip speed of 1,000 feet per minute by 4 to 6 operators to produce 1,000,000 base boxes per year (one base box = 218 square feet).

The basic steps in this process are:

1. Electrocleaning
2. Rinsing
3. Electropickling and chemical pickling
4. Washing and scrubbing
5. Plating
6. Drying
7. Brightening

The development of such a process required the skill of metallurgists and mechanical engineers for guiding the strip, of electroplaters for the chemistry of the plating, and of electrical engineers for many unique electrical controls.

At the beginning end of all of the electrolytic tinning lines, two coil holders, cropping shears and a welding unit are arranged. As the steel on one coil holder is being run through the machine, another coil is brought into position and made ready to be welded to the one being plated. In order to provide sufficient time for the welding operation and also to eliminate the necessity of slowing down the line, a looping pit or multiple-strand looper is incorporated in the line. The slack provided by the looper is sufficient for the welding operation, which requires only 4 to 5 seconds and permits continuous high-speed operation on the line.

In high-speed lines of this type, the design of the pinch rolls and tension bridles is important to maintain the proper pull and tension on the strip. Automatic connections are made, including automatic changes in current density to correct for changes in strip speed.

After the preparatory operations are completed, the strip is ready for plating. The plating is carried out either by an acid or by an alkaline process and with the strip traveling either vertically or horizontally through the electrolyte. Each of the processes and methods has its own advantages and disadvantages, but it is not within the scope of this book to go into details of the subject.

In the plating tanks, the solution is sometimes pumped against the strip. This movement of solution and the speed of the strip make it possible to deposit the tin at high current densities and, consequently, at a high plating rate. In fact, such high amounts of current are used in the conductor rolls in some installations that the rolls must be water cooled and in addition special precautions must be taken to prevent arcing which causes defects in the strip.

The high strip speed employed by most lines is accompanied by high drag-out rates, but efficient drag-out recovery systems reduce the metal loss to a minimum.

After leaving the plating unit, the strip is washed and dried and then usually given a brightening treatment. The brightening step is carried out either by scratch brushing with nickel-silver brushes or by actually melting and flowing the tin over the surface of the steel. At present, practically all of the tinned strip is brightened by melting. The strip is heated to a temperature slightly over 450°F. by gas flame, by immersing in hot oil, or by electrical means. Two electrical heating methods have met with wide favor: 1. by conduction, i.e., by passing an electrical current through the strip by contact rolls or 2. by induction, i.e., by passing the strip through a high frequency field of magnetic flux.

After being flow-brightened, the strip is usually given a chromate treatment after which it is either recoiled or cut to size by a flying shear and then automatically separated into prime stock and rejects.

Continuous plating, in some cases using the same equipment, has been applied to electrogalvanizing [3] and plating of steel wire.[4]

Fig. 26 shows a continuous strip-plating machine in operation.

FIGURE 26
Continuous Strip-Plating Machine (Courtesy of the Wean Engineering Company)

REFERENCES

1. K. W. Brighton, *Trans. Electrochem. Soc.,* **84,** 227 (1943).
2. C. Frenkel, *J. Electrodepositors' Tech. Soc.,* **21,** 129 (1946).

3. E. H. Lyons, *Trans. Electrochem. Soc.,* **84,** 279 (1943).
4. J. H. Conolly and R. Rimbach, *Trans. Electrochem. Soc.,* **84,** 293 (1943).

CHAPTER 23

Applications of Electroplating

A consideration of past and present uses of electroplating may lead to the development of new applications of this versatile process.

Ornamental Electroplating

The early applications of electroplating were primarily for ornamental purposes and even today, the majority of electroplates are applied essentially for decoration.

More than 100 years ago, silver was electrodeposited on metals as a cheap method for producing silver-clad objects. These silver clad objects were valued because of the pleasant appearance of the plate and people have always surrounded themselves with objects that are pleasant to look at. The silver-plated objects were either partially useful and partially ornamental, such as tableware, or they were wholly ornamental, such as jewelry.

Today, we find electroplates applied for ornamental purposes on tableware, hardware, electrical fixtures, automobile trim and a host of other objects. In fact, it is almost impossible to go through a day without using at least one electroplated article.

The plates on inexpensive objects such as paper clips and safety pins must be produced at very low cost. In cases such as these, electroplating must produce deposits of satisfactory appearance, but economy has entered into the production.

Economic Applications

It is difficult to separate economy from ornamentation since it is often necessary to impart a pleasing appearance to an object at a very low cost. Even for the more expensive objects such as tableware, electroplates can be produced far more cheaply than sterling silver and yet the plate can be produced heavy enough to last a lifetime.

Some applications of electroplating, however, are for purely economic purposes. Plating that is applied almost entirely for corrosion resistance is in this class. For example, zinc, lead or tin may be applied to sheet metal to insure sufficient life under conditions of exposure to moisture, weathering or corrosive liquids.

Plating is often applied to increase the life of a part exposed to wear. In this case, the application is economic, but the use is fundamentally an engineering use.

Engineering Applications

Engineering applications are those in which a better job is done by plating than by some other means. Plating in this class is not widely used, but there are a number of important applications. These applications include salvage of improperly machined parts, build-up of worn parts, production of wear-resistant surfaces, bearings, electrical contacts, electroformed parts and selective plating prior to carburizing.

Research

Electroplating is a convenient research tool. High-purity metal can be produced by this means and used as a basis for the study of a pure metal and for studies in diffusion, resistivity and corrosion. The process is unique when compared to ordinary hot methods in that thin, uniform deposits can be produced at low temperatures. Deposits can be readily produced with little equipment, and new uses suggest themselves from consideration of the major current applications.

The following listing is of well-known applications, but does not include hundreds of minor uses that take advantage of the special properties of thin, uniform deposits.

Table 10

Major Applications of Electroplating

For	*Electroplate*
Automobile trim	Chromium (with undercoat of nickel and copper)
Battery parts	Lead
	Lead-tin
Bearings	Silver and silver alloys
	Lead alloys
	Tin alloys
Break-in (pistons)	Tin
Build-up of worn parts	Chromium
	Nickel
	Iron
Containers (food)	Tin
(chemicals and oil)	Lead and lead-tin
Corrosion resistance (weather)	Zinc
	Lead
	Cadmium
Corrosion resistance (chemical)	Specific for the chemical
Corrosion resistance (oil)	Lead and lead-tin
	Cadmium
	Copper-base alloys
	Nickel
Corrosion (voltaic-couple)	Cadmium (over dissimilar metals)
Eating utensils	Silver
	Nickel
Electrical contacts	Silver
	Silver-rhodium
	Gold
Electroforming (general)	Copper (acid)
	Nickel
	Iron
Electroforming (sheet)	Copper
	Nickel
Electroforming (tubes)	Nickel
Electroforming (screen)	Nickel
	Copper
Electronic equipment	Silver
	Silver and rhodium

TABLE 10 (*Continued*)

For	*Electroplate*
Electrotypes	Copper (acid)
Floats	Acid copper (electroformed)
Gun barrels	Chromium
Laboratory equipment	Silver
	Gold
	Nickel (copper undercoat)
Machines (food processing)	Tin
	Nickel
Machines (laundries)	Cadmium
Moisture resistance (general)	Zinc
(marine)	Cadmium
(steel in concrete)	Cadmium
(structural)	Lead
(washing equipment)	Cadmium
Molds (lining)	Chromium
Musical instruments	Silver
	Gold
Nuts and bolts	Lead
	Zinc
	Cadmium
	Nickel
	Chromium
Ornamental	Silver
	Gold
	Rhodium
	Nickel
	Chromium
Pins, paper clips, etc.	Tin
	Copper
Plumbing fixtures	Chromium (nickel and copper undercoat)
Powdered metal	Copper (acid)
	Iron
Printing plates	Nickel
	Iron
Reflectors (general)	Cobalt
	Lacquer over silver
	Chromium

TABLE 10 (*Continued*)

For	*Electroplate*
	Speculum
	Anodized aluminum
Reflectors (infrared)	Gold
Refrigeration coils	Tin
Rubber adhesion	Brass
Salvage of worn parts (specific)	Depending on application
(general)	Chromium
	Nickel
	Iron
Selective carburizing	Copper
	Bronze
Sheet metal (food containers)	Tin
(roofing)	Lead
	Lead-tin alloy
(outdoor exposure)	Zinc
Strike plating baths (general)	Cyanide copper
(silver-plating)	Cyanide silver
(cadmium adhesion)	Acid cadmium
Tarnish resistance	Chromium
	Tin
	Gold
	Platinum
	Chromium (nickel and copper)
	Nickel (copper)
Wear (hardness and low coefficient of friction)	Chromium
(hardness and electrical conductivity)	Rhodium over silver
(limited wear)	Nickel
	Iron
(cylinder liners)	Porous chromium
Wire (steel)	Zinc
(copper)	Tin

CHAPTER 24

Plating Bath Troubles *

A plating process, to be acceptable, must insure continuous quality production by test, by observation or by analysis.

Experienced platers reduce the essential control methods for their plating baths to a minimum by adjustment of the anode area in order to maintain proper metal balance in solution, by periodic additions to the baths based on the amount of work processed and by observation of the appearance of the finished product. Zinc cyanide baths for example, have been successfully operated in this fashion for years with the aid of frequent on-the-spot analyses for total cyanide and an occasional complete analysis by an outside laboratory. Through correlation of these tests with production results, the plater adds facts to his experience which enable him to operate at maximum efficiency.

A less experienced plater would require frequent analyses of all the important bath ingredients as well as operation instructions explaining the meaning of the analyses and placing limits on all variables requiring control. However, by providing analytical facilities and specifying narrow bath limits in the operating instructions, it is possible for a plating engineer to set up a process that can be operated successfully by men with relatively little training. Although there is considerable difference in the methods of these two kinds of operation, the same quality of plate can be obtained in either case and both processes have proved satisfactory.

All analyses may be seldom eliminated from a process, but

* Originally published in *Metal Finishing*.

in some cases, such as that of the zinc cyanide bath, a plater may never analyze for free caustic and yet he may control it by adding caustic in a definite ratio to the cyanide added, when the cyanide is controlled by analysis.

In the best-controlled process, inferior plating will occasionally be obtained as the result of a variable that is not or cannot be controlled or foreseen.

A condition such as a porous basis metal may give rise to a blistered plate. Partially decomposed cooling oil, due to a change in rolling practice at the steel mill, may produce a surface that cannot be cleaned by established practice.[1] Poor plate adhesion from freshly prepared baths may be the result of impurities accumulated in the chemical processing of the plating salts. A rough plate may be caused by suspended solids carried over from the anode sludge, from dirt in the air or from precipitation of compounds in the bath. The deposit may blister or fail to cover due to improper or inadequate cleaning.

The part to be plated may not be covered continuously due to improper anode spacing or arrangement, or to poor rack design. The plate may suddenly develop spots that can be due to such a variety of reasons that they may only be corrected through specific knowledge of previous troubles of the same sort or by extensive experimentation.

These common troubles are not readily measurable but are generally detected and controlled by inspection of the finished product.

In case of unexpected troubles, a check may first be made of all controllable limits. If all limits are under control, a check may be made on the cleaning line, as a variety of troubles are caused by poor cleaning. A variable that does not show up in analysis, such as excessive oil in the cleaner, may easily be the offender. It is usually possible to by-pass the cleaner tank, by hand cleaning a few pieces with solvent, followed by a mild abrasive paste, and confirm or eliminate trouble at this point or in the pickle by a similar procedure. If the plate is rough, filtration is in order and a rapid check may be made on a small portion of the bath on a laboratory scale before and after filtering.

If impurities in the bath are suspected, a purification step may be tried. Low-current-density electrolysis, such as is commonly used for nickel baths,[2] may remove metal impurities. Other chemical purification steps can be taken depending on the bath and the impurity suspected. In some cases, it is customary to analyze for and describe limits for tolerance of known impurities, such as iron in a nickel bath.[3]

These general troubles are well known to platers, but the specific troubles found through misfortune and experiment are the offenders that interest us most. Some indications of these are given in the following tables although, of course, it is impossible to know or record a complete list.

The bath troubles listed are taken from the literature,[4] from private sources and from experience. In most cases, the troubles were listed only where the same effect was agreed on by more than one source. In many cases, no remedy was suggested. The tables of general troubles do not apply to any particular bath, but are common in many cases.

General Bath Troubles

Off Color Deposits

1. Analysis: Make a complete analysis of the bath and adjust all ingredients to the proper chemical limits.

2. Thin Deposits: Plate a heavy deposit to determine if coverage is adequate to produce the desired color.

3. Metal Impurities: Purify by low-current-density electrolysis or by chemical treatment. Remove the solution from the tank, examine for parts dropped in the tank, for dirt and for electrical faults. Clean the tank and filter the solution into it.

Blistered Deposits or Poor Adhesion

1. Cleaning and Pickling: By-pass these steps by using separate tanks or hand cleaning. Examine the part for oil or grease not removed by ordinary cleaning methods.

2. Fresh Baths: Very often, deposits from a fresh bath will have poor adhesion, but this trouble will gradually disappear if the bath is electrolyzed for a period of time to remove impurities sometimes present in commercial salts.

3. Basis Metal: Examine the basis metal for cracks or holes that may trap solution in the cleaning steps and prevent proper plating.

4. Strike Baths: Very often, the addition of a strike step will overcome poor adhesion. A Rochelle copper bath is often used for this purpose.

Rough Deposits

1. Suspended Material: Suspended solids in a bath often cause rough deposits. These may be eliminated by filtering.

2. Poor Anode Corrosion: If the anode corrodes so that particles of metal form a loose film on the surface of the anode, it is quite common for these particles to be carried to the cathode and cause a rough deposit. Filter the bath and bag the anodes.

3. Dirt on the Work: Often particles, such as grinding compound, small rubber or other particles, e.g., those used in previous process steps, are not removed by the ordinary cleaning steps. Hand clean the work with a mild abrasive such as magnesium oxide until no discoloration is obtained after wiping with a clean cloth.

4. Overplating: Some baths, such as a copper cyanide bath will invariably produce rough deposits for thicknesses of the order of 0.003 inch or greater.

5. High Current Density: Many baths, particularly the acid types, will plate roughly if operated at too high a current density. Cathode agitation permits the use of higher current densities.

6. Low Metal Content: Operation at normal current densities with low metal content will give an effect similar to operation at high current densities with normal metal content. Agitation will help, but addition of metal is more satisfactory.

7. Agitation: If agitation is used during plating, such as in

the high-efficiency copper cyanide process, frequent or continuous filtration is necessary, since impurities that would normally settle to the bottom of the tank will be suspended by the agitation.

8. Impurities: If impurities are suspected, they can only be found by previous experience or investigation. The important fact is that impurities have to enter the bath from some direct contact source so that drag-in, chemicals used, water used, anodes, the plating atmosphere and the work itself may all be suspected. Very often, the parts plated, such as zinc-base die castings, may corrode to some extent before they are covered and eventually cause trouble. Racks that are used for plating in more than one type of bath may easily cause contamination.

Spotty Deposits

Spotty deposits are sometimes caused by poor cleaning and occasionally by pick-up of oil that has accumulated on the surface of the cleaner.

Poor Anode Corrosion

1. Impurities in Anode: Specify anode purity.
2. High Anode-Current Density: Increase the anode area.
3. Impurities in the Bath: Purify the bath and analyze for excessive accumulation of chemicals such as carbonate in alkaline baths.
4. High Metal Content: If the metal content is high, the concentration of metal near the anode readily becomes excessive on electrolysis.

Low Free Acid

If the acid is low or a buffering acid is low in a high-pH acid bath, the free acid is readily used up at the anode.

Low Free Cyanide

It has the same effect as low free acid.

Low Cathode Efficiency

1. Low Metal Content: If the metal content is low, the ability for hydrogen to evolve is greatly increased.

2. High Free Cyanide: Excessive amounts of cyanide will lower the cathode efficiency in many cyanide baths.

3. Impurities: Many cases of lowering of cathode efficiency by accumulation of impurities are known, but no general rule can be stated.

In some cases, the accumulation of ammonia in cyanide baths, due to decomposition of the cyanide, will lower the cathode efficiency. This can be remedied temporarily by heating the bath to drive off ammonia.

High Metal Content

Excessive anode area often leads to the build-up of metal in the bath.

Plate Distribution

1. Recesses: In general, the acid baths are poor in throwing power, and the alkaline baths are good. In case difficulty is experienced in plating in recesses, the location of the anode and the design of the rack should be considered. By allowing the anode to extend into a recess, better covering may be obtained.

2. Corners and Edges: A bath that will not plate in recesses will also be a bath that will build up a deposit on corners and edges. This effect may be partially overcome by robbing the high-current-density areas through the use of auxiliary cathodes near these areas or by the use of nonconducting material between this area and the anode to shadow the area. Each case is a separate problem that must be solved by experience and experiment although some general theoretical principles are available.[5]

Brass

The following specific troubles are not complete and do not

apply in every case, but they indicate what may occasionally occur in each typical bath.

Typical Bath Composition: [6]

	g./l.	*oz./gal.*
Copper Cyanide	30.0	4.00
Zinc Cyanide	9.4	1.25
Sodium Cyanide	56.3	7.50
Sodium Carbonate	30.0	4.00

Since brass is an alloy, the effect of a given variable cannot be stated with surety since copper may tend to plate more readily from one bath and zinc from another. The metal that tends to deposit more readily is called the noble metal by the alloy plater and its deposition may be decreased by increasing the current density. For the typical bath composition given, an increase in current density will decrease the percentage of zinc in the deposit, but for another composition, the opposite effect may be attained.[7]

Pink or Red Deposits

This condition is a result of the deposit containing too high a percentage of either zinc or copper. If the trouble is caused by high copper content, it may be corrected in a number of ways depending upon the condition of the bath.

1. By lowering the bath temperature.
2. By adding caustic to raise the pH.
3. By lowering the current density.
4. By adding zinc cyanide.
5. By adding both zinc cyanide and copper cyanide to lower the free cyanide content.

Reddish deposits caused by high zinc content may be remedied as follows:

1. By raising the bath temperature.
2. By adding sodium bicarbonate to lower the pH.
3. By raising the current density.
4. By adding copper cyanide.

5. By adding alkali cyanide to increase the free cyanide content.

Off-Color Deposits

Addition of small amount of ammonia or ammonium salts may temporarily correct this condition.

White or Gray Deposits

Excessive amounts of arsenic or nickel in the bath will cause this trouble. Electrolysis is the most satisfactory method of removing or reducing the content of the contaminating metals.

Dull-Dark Deposits

This means lead contamination which is best corrected by electrolysis.

Poor Anode Corrosion

The presence of tartrates or citrates and a low free cyanide content result in poor anode corrosion.

Cadmium Cyanide

Typical Bath Composition:[8]

	g./l.	*oz./gal.*
Cadmium Oxide	23– 39	3 – 5.2
Cadmium Metal	19– 34	2.5– 4.5
Sodium Cyanide	86–131	11.5–17.5

The cadmium cyanide bath is susceptible to deleterious effects due to metallic impurities because of the tendency of many metals to deposit readily from this electrolyte. This means that these impurities must be kept from the bath by using high-purity anodes,[9] and it also means that the impurities may usually be removed by electrolysis at low current density or by the addition of cadmium sponge.

Most cadmium baths are operated to give bright deposits through the use of addition agents.

Dull Deposit

1. High Metal Content: Electrolyze with steel anodes or add cyanide to reduce the concentration of metal ions.
2. Low Cyanide: Add sodium cyanide.
3. High Carbonate: Remove carbonate by freezing or precipitation as calcium carbonate.[10]
4. High Caustic: Add cadmium cyanide in place of cadmium oxide.

Burnt Deposit

This condition is caused by poor plating in high-current-density areas and, in general, is due to conditions causing a low cathode-current density. Usually, it is caused by a low metal content.

Nonuniform Deposit

A small amount of sodium carbonate is said to be beneficial to operation of the bath. Hull [11] recommends limits of 2 to 6 oz./gal. Low carbonate may cause nonuniform deposits.

Slate-Colored or Streaked Deposits

Lead, copper, silver, antimony, thallium and tin may cause inferior deposits. These metals may be removed by electrolysis or by precipitation with cadmium sponge followed by filtering of the bath.

Pitted Deposits

In this case, the caustic content should be checked. If it is low, it may cause pitting. If this does not correct the trouble, the cleaning and pickling steps may require improvement.

Brittle Deposit

Hydrogen embrittlement due to overpickling.

Polarized Anodes

At high anode-current densities, the anodes become black, powdery and begin to evolve oxygen. This undesirable condition leads to difficulties in bath control and, in some cases, to a rough deposit from the following causes:

1. High Anode Current Density: Increase the anode area.
2. Low Caustic: Add caustic soda.
3. Low Cyanide: Add sodium cyanide.
4. High Carbonate: Remove carbonate by crystallization or precipitation.

Uneven Dissolution of Anodes

This condition may be caused by the accumulation of insoluble impurities on the surface of the anode and can only be overcome by the use of high-purity anodes.

Immersion Plating on Anodes

Contamination of the bath by metals nobler than cadmium may result in immersion plating on the anodes. These metals may be removed by electrolysis at low current density or by precipitation with cadmium sponge.

High Bath Voltage

1. Polarized Anodes: See foregoing.
2. Poor Bath Conductivity: Usually due to low caustic content of the bath.

Low Cathode Efficiency

1. Low Metal: Add cadmium oxide to the bath.
2. High Cyanide: Add cadmium oxide to the bath.

Chromium

Typical Bath Composition: [12]

	g./l.	*oz./gal.*
Chromic Acid	250.0	33.30
Sulfuric Acid	2.5	0.33

The chromic acid plating bath is notable for its poor throwing power, and parts to be plated should be arranged in the bath or on a plating rack so that the best possible current distribution is obtained. If this fact is kept in mind along with the facts that the solution is highly corrosive and more current is required than in most plating baths, very little trouble will be experienced.

Narrow Plating Range

This is caused by either too high or too low a CrO_3 to SO_4 ratio. If the ratio is appreciably low, the deposit will assume a greenish-yellow tinge while if the ratio is high, additions of catalyst acid radical will result in a further decrease in the plating range.

Thin Deposits—Low Efficiency

Both temperature and CrO_3 to SO_4 ratio are associated with plating rate and coverage. At a constant current density, the plating rate decreases as the temperature is increased.

Milky-Streaked Deposits

This condition is usually caused by too low a current density.

Dull-Matte Deposits

A low operating temperature will give this type of plate.

Burned-Nodular Deposits

The current density is too high.

Poor Bath Conductivity

This indicates bath contaminated with iron, copper, zinc, etc.

Trivalent Chromium Build-up

The use of iron or steel anodes will result in a higher equilibrium concentration of trivalent chromium than when the conventional lead or lead-alloy anodes are used. This build-up can also be caused by too small an anode area, using lead or lead alloy anodes, as an insufficient surface area of lead peroxide is present to reoxidize the chromium which is reduced at the cathode.

Acid Copper

Typical Bath Composition:[13]

	g./l.	*oz./gal.*
Copper Sulfate	150–250	20–33
Sulfuric Acid	45–100	6–13

The acid copper bath is easy to operate and control, but is not likely to replace the copper cyanide bath because no commercial bath yielding bright deposits is available and because a copper strike is required to obtain adhesion on basis metals, such as steel or zinc alloys.

Hard Deposits[14]

1. Low Metal Content: Add copper sulfate.
2. High Acid: Add copper carbonate.
3. High Current Density.
4. Low Temperature.

Treed Deposits

This condition indicates insufficient addition agent. Add glue.

Nonuniform Deposits

Deposits of this type are sometimes caused by the presence of organic material which is usually colloidal and, if troublesome, may be removed by treatment with activated carbon, followed by filtration.

Excessive Copper Sulfate in the Bath

This condition indicates that the anode area is too large and it should be reduced to approximate the cathode area.

Rough Deposits

1. Suspended Solids.
2. Anode Particles: Use rolled copper anodes.

Rochelle Copper

Typical Bath Composition:[15]

	g./l.	*oz./gal.*
Copper Cyanide	26	3.5
Sodium Cyanide	35	4.6
Sodium Carbonate	30	4.0
Rochelle Salt	45	6.0

This type of bath or similar low-efficiency copper cyanide baths are frequently used as an undercoat for other deposits to promote adhesion. It may, therefore, be regarded as a process bath or one that is used as a step in a plating process where another metal is deposited over the copper, but where the process would fail in many cases if this step were not included.

Dull Deposits

This condition is usually caused by one or more of the following:

1. Low Free Cyanide Content.
2. Low Rochelle Salt Content.
3. High Carbonate Content: Can be lowered by either crystallization or precipitation.

Burnt Deposits

Baths containing very high free cyanide will give this result under normal operating conditions. Immediate correction is obtained by adding copper cyanide and then replacing some of the steel anodes with copper anodes in order to prevent metal depletion with consequent free cyanide build-up.

Nonuniform Deposits

Nonuniformity is usually caused by:

1. Low Free Cyanide.
2. Low pH.

Rough Deposits

1. Suspended Material in the Bath.
2. High pH.

Brassy Irregular Deposits

Zinc Contamination: The zinc can be removed by electrolysis or by precipitation with sodium sulfide, although an excess of sulfide may cause nonuniform deposits.

Dark Deposits

Lead Contamination: Very small amounts of lead act as brighteners or crystal refining agents while larger amounts cause the deposit to become dark and smutty. The lead can be removed by precipitation as sulfide or by electrolysis.

Low Cathode Efficiency

A low over-all concentration, high Rochelle salt content and very high free cyanide all tend to lower the cathode efficiency.

Polarized Anodes

This may be overcome by the use of iron and copper anodes as the iron tends to depolarize the copper. A very low carbonate content also tends to favor this condition.

Poor Anode Dissolution

One or more of the following may cause this trouble:

1. High pH.
2. High Carbonate.
3. High Anode-Current Density.

Iron Anodes Corroded

The carbonate content of the bath should be maintained within specified limits to prevent corrosion of the iron anodes.

Bluish Cast of the Solution

Cupric ions give this color indicating little or no free cyanide.

High-Efficiency Sodium Cyanide Copper

Typical Bath Composition:[16]

	g./l.	*oz./gal.*
Copper Cyanide	120.0	16.0
Sodium Cyanide	135.0	18.0
Sodium Hydroxide	30.0	4.0
Brightener	15.0	2.0
Antipit Agent	1.5	0.2

Dull Reddish Deposits

A high free cyanide content will cause this condition in the higher-current-density areas and low brightener and antipit agent contents, high carbonate content and too slow cathode agitation may also be contributing factors. Low free cyanide content will tend to shift the bright plating range toward the higher-current-density area, while low caustic content will result in a narrow bright-plating range at low-current-density areas.

Pitted Deposits

Pitting occurs in high-current-density areas when the antipit agent content is low.

Water Breaks

High antipit agent content causes water breaks.

Dark Smutty Deposits

This is a sign of lead contamination which should also become apparent by a lead film on the anodes.

Bronze to Brass Color

Zinc contamination.

Mottled, Nonuniform Deposits

Organic contamination which may be corrected by treatment with activated carbon.

Poor Coverage; Peeling

Chromium contamination which may be corrected by reducing the hexavalent chromium to trivalent chromium with sodium hydrosulfite.

Polarized Anodes

Too high an anode-current density causes the anodes to polarize with a black insoluble film of copper oxide while too low a current density results in the formation of a loose sludge which tends to roughten the cathode. Operation at low temperatures also causes anode polarization.

High-Efficiency Potassium Copper Cyanide

Typical Bath Composition: [26]

	g./l.	
Copper Cyanide	56.0	7.5 oz./gal.
Free Potassium Cyanide	3.8	0.5 oz./gal.
Potassium Hydroxide	41.0	5.5 oz./gal.
Antipit Agent		0.3 fl. oz./gal.

The high-efficiency copper cyanide baths, known as high-speed copper, will plate brighter deposits than the Rochelle bath and bath control is not so critical since both the anode and the cathode efficiencies are 100 per cent. The plating rate is high, and smooth, bright deposits may be obtained up to 0.002 inch in thickness. The principal disadvantages of these baths are that a copper strike must be used to prevent immersion plating and the cathode must be moved relatively to the solution during plating.

The potassium bath can be operated at a lower metal content than the sodium bath and with less bath decomposition.

Dull Deposits

Dull deposits from this bath are regarded as those that have the typical color of a Rochelle bath.

1. High Current Density: Do not exceed 60 amperes per square foot. If higher current densities are required use a higher metal content.
2. High Free Cyanide.
3. Low Temperature.
4. Low Cathode Movement Rates: Cathode movement may be supplemented by solution agitation if continuous filtration is used.

Burnt Deposits

In general, the conditions that cause dull deposits will cause burnt deposits if carried to an extreme.

Rough Deposits

1. Suspended Material in the Bath.
2. Sludge from the Anode: Use electrolytic copper anodes. Increase the anode-current density.

Off-Color Deposits

1. Zinc: Zinc will cause a brassy deposit.
2. Lead: Lead will cause a dark rough deposit.

Anode Polarization

1. Excess Carbonate.
2. Low Free Cyanide.
3. High Anode-Current Density.
4. Low Temperature.

Low Cathode Efficiency

1. Low Metal Content.

2. High Free Cyanide.
3. Low Temperature.

Pitted Deposits

1. Low Cathode Efficiency.
2. Low Antipit Agent.

Low Conductivity

Low Caustic Content.

Cyanide Decomposition

1. Low Anode Efficiency: Increase the anode area and check caustic content.
2. High Temperature.
3. High Free Cyanide.

Water Breaks

High Antipit Agent: If water will not form a continuous film over the cathode after plating and rinsing, treat the bath with activated carbon. Cathodic cleaning after plating may serve as a temporary remedy.

Gold Cyanide

Typical Bath Composition: [17]

	g./l.	*oz./gal.*
Metallic Gold	2.1	0.25
Potassium Cyanide	15.0	2.00
Dipotassium Phosphate	4.0	0.50

The most significant thing that can be said about gold baths is that there has been relatively little published about them as compared to the other well-known baths.

Thin Deposits—Low Cathode Efficiency

1. Low Metal Content.
2. Low Free Cyanide.

3. Low Temperature.
4. High Current Density.

Color of Deposits

1. Pink or Red: Copper.
2. Green: Silver and Cadmium.
3. White: Nickel and Zinc.
4. Black: Lead.

High Metal Content

This is due to excessive anode area.

Off-Color Deposits

1. Impurities.
2. Deposit Too Thin.

Acid Lead

Typical Bath Composition:[18]

	g./l.	*oz./gal.*
Lead Fluoborate	105	14
Free Fluoboric Acid	40	5.3
Glue	0.2–5	0.03–0.7

Low Lead Content

1. Spongy Deposit at High Current Density.
2. Poor Covering Power at Low Current Density.

High Lead Content

No apparent effect.

Addition Agent

1. Crystalline Plate: Add gelatin or glue.

2. Brittle Plate: It is due to excessive addition agent. Treat with activated carbon.

3. Control by Experience.

Acid Iron

Typical Bath Composition: [19]

	g./l.	*oz./gal.*
Ferrous Chloride	400–500	54–67

The ferrous chloride bath is operated at high temperature with periodic additions of hydrochloric acid to maintain an acidity of about 0.01 normal. Little trouble is experienced with metallic impurities.

Brittle Deposits

1. Low Temperature.
2. Organic Impurities: Treat with activated carbon.

Pitted Deposits

Ferric Iron: Add excess acid to reduce ferric iron by chemical action on the anodes.

Rough Deposits

1. Suspended Solids in the Bath. Filter.
2. Particles from the Anodes. Use ingot iron anodes and bag if necessary.

Silver Cyanide

Typical Bath Composition: [20]

	g./l.	*oz./gal.*
Silver (Metal)	25–33	3.3– 4.4
Free Potassium Cyanide	30–45	4– 6
Potassium Carbonate	30–90	4–12

Burnt Deposits

Dark or rough deposits in high-current-density areas may be avoided by agitation, increasing the metal content, increasing the temperature or the use of addition agent.

Rough Deposits

1. Suspended Solids.
2. Particles from the Anodes.
3. Metallic Impurities: Including lead, bismuth, manganese, tellurium, selenium, sulfur, antimony. If the specific trouble is not known, treat the bath with activated carbon, bag the anodes, filter and electrolyze at low current density.

Polarized Anodes

1. High Current Density.
2. High Carbonate.
3. Low Cyanide.
4. High Metal Content.

Alkaline Tin

Typical Bath Composition: [21]

	g./l.	*oz./gal.*
Sodium Stannate	90.0	12
Caustic Soda	7.5	1
Sodium Acetate	15.0	2
Hydrogen Peroxide (100 vol.)	0.5	1/16

Dark Spongy Deposits

This may be the result of contamination by lead, antimony or arsenic, which may be corrected by electrolysis or, it may be due to the presence of stannite ions which may be oxidized to stannate ions by the addition of hydrogen peroxide or sodium perborate. The conditions leading to the formation of stannite ions are: 1. too low an anode current density and 2. improper anode current density-caustic soda content relationship.

Low Cathode Efficiency

This can result from:

1. Too high cathode-current density.
2. Too low caustic soda content.

3. Too low operating temperature.
4. Too low over-all concentration.

Excessive Insoluble Salt Formation

This not too serious problem occurs when the alkalinity of the bath is too low and carbon dioxide is absorbed from the air or when the water used is hard.

Poor Throwing Power

Throwing power is improved by the following:

1. Increasing the metal concentration.
2. Increasing the caustic content.
3. Increasing the carbonate content.
4. Addition of sodium acetate.

Anodes

At optimum operating conditions, the anodes are coated with a greenish-yellow film. Grayish anodes indicate too low current density while black anodes indicate to high current density.

Acid Zinc

Typical Bath Composition: [22]

	g./l.	*oz./gal.*
Zinc Sulfate Crystals	240	32
Ammonium Chloride	15	2
Aluminum Sulfate	30	4
Licorice	1	0.13

Rough or Pitted Deposits

Suspended Material. Filter the solution and bag the anodes.

Blistered Deposits

Improve the cleaning cycle.

Low Cathode Efficiency

1. Low Metal.
2. Excess Addition Agent.

Metallic Impurities

1. Cadmium. Remove with sponge zinc.
2. Iron. Oxidize with peroxide, neutralize the bath with zinc metal and filter off the ferric hydroxide.

Zinc Cyanide

Typical Bath Composition: [23]

	g./l.	*oz./gal.*
Zinc Cyanide	60	8
Sodium Cyanide	23	3
Sodium Hydroxide	53	7

Zinc plating is extensively used to prevent rusting of iron and steel.

In recent years, processes have been developed for preparing bright plate through the use of addition agents.

Dull Deposits

1. Low NaCN to Zn Ratio: Increase NaCN or reduce Zn.
2. High Caustic Soda.
3. Metallic Impurities: Lead, copper and cadmium can be removed by treatment with zinc dust followed by filtering to prevent resolution of the precipitate. Lead may be removed by treatment with sodium sulfide.

Blistered Deposits

1. Poor Pickling.
2. High NaCN to Zn Ratio.
3. Metallic Impurities.

Crystalline Deposits

High Caustic Soda.

Excessive Anode Polarization

Low Caustic Soda.

Excessive Gassing—Low Cathode Efficiency

1. Low Metal.
2. High NaCN to Zn Ratio.

Bright Deposits

1. Addition Agent Control. Can be done by careful observation of the deposit or by plating tests.[24]
2. Careful Control of Chemical Limits.
3. Control of Metallic Impurities in Bath.

Nickel

Typical Bath Composition:[25]

Watts Bath

	g./l.	*oz./gal.*
Nickel Sulfate	240	32
Nickel Chloride	45	6
Boric Acid	30	4

The Watts' type bath operated at a pH range of 1.5 to 3.0 or 4.5 to 5.6 has met with wide favor in the industry and the following discussion will be limited to this bath.

Addition-Agent Control

The bright-nickel baths used for protective and decorative plating contain one or more addition agents to refine the grain size and brighten the deposit. Most of these brighteners are proprietary compounds, and the methods for their control are given by their suppliers.

pH Range

The choice of pH range is governed by the type of work being processed or by the final results desired. The advantages of operating at a high or at a low pH are listed as follows:

a. Low pH:
1. Wide Current Density Plating Range.
2. High Permissible Current Density.

3. Less Tendency Toward Rough Deposits Caused by Metallic Impurities.
4. Softer Deposits.

b. High pH:
1. Superior Throwing Power.
2. Superior Covering Power.
3. High Cathode Efficiency.
4. Harder Deposits.

Poor Throwing Power

a. Low-pH Bath:
1. Temperature Too Low.
2. Low Current Density.
3. Low Metal Content.
4. Iron Contamination: Remove by first treating with permanganate, then with nickel hydroxide or nickel carbonate.
5. Excessive Amount of Hydrogen Peroxide.
6. Copper or Zinc Contamination.
7. Organic Contamination Reduces Covering Power.

b. High-pH Bath:
1. High Current Density.
2. Excessive Amount of Peroxide.
3. Contamination by Other Metals.

Low Cathode Efficiency

1. Low Chloride Content Reduces the Cathode Efficiency at Low Current Densities.
2. Too High Current Density.
3. Too Low Temperature.
4. Low Over-All Metal Concentration.
5. Too Low pH.
6. Copper or Zinc Contamination.

Poor Conductivity

Low Over-All Metal Content.

Burnt Deposits

1. Low Metal Content.
2. Low Operating Temperature.
3. Low Chloride Content.
4. Excessive Amount of Hydrogen Peroxide.
5. Too High Current Density.

Pitted Deposits

1. Little or No Boric Acid.
2. Metallic Impurities in the Bath.
3. Need for Wetting Agents.
4. Need for Hydrogen Peroxide Treatment if No Organic Addition Agents Are Present.

Rough Deposits

a. Bath Composition:

1. Inclusions of basic materials because of high pH, absence of boric acid, or high current density.
2. Low Chloride. Chloride tends to improve smoothness of the deposit.

b. Contaminants Removed by Filtration:

1. Anode Particles and Sludge.
2. Dust Particles.
3. Ferric Hydroxide, if the operating pH is high enough to cause its precipitation.
4. Insoluble material contained in plating salts.

Soft Deposits

1. Too High Boric Acid Content.
2. Too High Operating Temperature.
3. Low pH.

Hard, Brittle Deposits

1. Excessive Hydrogen Peroxide.
2. Absence of Boric Acid.

3. Alkali or Ammonium Salt Contamination.
4. Iron Contamination.

Abnormal pH Variation

1. Low Boric Acid.
2. Poor Anode Corrosion Caused by Low Chloride, Excessive Anode-Current Density, High pH or Impurities in the Anodes.

Off-Color Deposits

1. Copper and Zinc Contamination Especially in Low-Current-Density Areas: These are most satisfactorily removed by electrolysis at extremely low current density.

Anodes Polarized or Poorly Corroded

1. Too High Anode-Current.
2. Too Low Chloride Content.
3. Impurities in the Anodes.
4. Too High pH.

Conclusions

For successful continuous operation of plating baths one must be prepared to deal with trouble. Specific troubles are so numerous that the only safeguard against them is a knowledge of the tools and methods used to investigate the behavior of the anode, the electrolyte, the quality of the deposit and the endurance of the product.

The literature answers most routine questions, but extensive literature references and lively discussion over a paper on modern nickel baths [25] indicate that it is not feasible to find all the answers in the literature. The answer to the problem is to study the process and the product and to learn how to control each operation or property and how to use and interpret significant tests in the event of trouble.

References

1. E. H. Lyons, *Trans. Electrochem. Soc.,* **80,** 367 (1941).
2. Weisberg, *Metal Ind.* (N. Y.), **34,** 451 (1937).
3. Madsen, *Trans. Electrochem. Soc.,* **45,** 249 (1924).
4. Many of the troubles are described in volume 80 of the *Trans. Electrochem. Soc.* and in the *Plating and Finishing Guidebook.*
5. C. Kasper, *Trans. Electrochem. Soc.,* **77,** 353 (1940).
6. G. B. Hogaboom, Jr. and N. Hall, *Plating and Finishing Guidebook,* 1945, p. 13.
7. Ferguson and Sturdevant, *Trans. Electrochem. Soc.,* **38,** 167 (1920).
8. G. Soderberg and L. R. Westbrook, *Trans. Electrochem. Soc.,* **80,** 431 (1941).
9. G. Soderberg, *Monthly Rev., Am. Electroplaters' Soc.,* **19,** 9 (Feb. 1932).
10. R. O. Hull, U. S. Patent 2,164,924.
11. R. O. Hull, *Plating and Finishing Guidebook,* 1945, p. 17.
12. G. Dubpernell, *Trans. Electrochem. Soc.,* **80,** 589 (1941).
13. J. H. Winkler, *Trans. Electrochem. Soc.,* **80,** 521 (1941).
14. Soft deposits are not always desirable.
15. A. K. Graham and H. J. Read, *Trans. Electrochem. Soc.,* **80,** 341 (1940).
16. H. L. Benner and C. J. Wernlund, *Trans. Electrochem. Soc.,* **80,** 355 (1940).
17. L. Weisberg and A. K. Graham, *Trans. Electrochem. Soc.,* **80,** 509 (1941).
18. F. C. Mathers, *Plating and Finishing Guidebook,* 1945, p. 71.
19. C. T. Thomas, *Trans. Electrochem. Soc.,* **80,** 499 (1941).
20. N. E. Promisel and D. Wood, *Trans. Electrochem. Soc.,* **80,** 459 (1941).
21. F. F. Oplinger and Fred Bauch, *Trans. Electrochem. Soc.,* **80,** 617 (1941).
22. W. Blum and G. B. Hogaboom, *Principles of Electroplating and Electroforming,* 1930, p. 316.
23. R. O. Hull and C. J. Wernlund, *Trans. Electrochem. Soc.,* **80,** 1941).
24. R. O. Hull, *Monthly Rev., Am. Electroplaters' Soc.,* **26,** 753 (1939).
25. W. L. Pinner, G. Soderberg and E. M. Baker, *Trans. Electrochem. Soc.,* **80,** 539 (1941).
26. H. Brandes, *Trans. Electrochem. Soc.,* **88,** Preprint 26 (1945).

CHAPTER 25

Analytical Methods for Plating Baths

Analytical methods are the foundation for plating-bath control. Although it is possible to go far in controlling many baths without resorting to analytical methods, it is not necessarily desirable to eliminate analysis. By the use of analytical control it is possible to hold a bath in limits, whereas if control were maintained purely by physical and plating tests, the bath would gradually drift away from the original limits. Eventually a point would be reached where radical changes in the bath would be required to restore it to the original limits. This usually results in temporary shutdown of the tank and loss of time and production.

Analyses are not made to get out of trouble but rather to stay out of trouble. By frequent analyses and continuous adjustment of chemicals, baths have been operated for years without trouble.

It is worth while to review analytical methods recommended for plating-bath control and it is worth while to take an occasional look at the standard analytical books. Plating methods are rapid and relatively inaccurate compared to standard routine analytical methods. This relative inaccuracy is allowable for the wide limits used for plating control; but it should be realized that, if accuracy is required, a better method is preferable. For instance, such accuracy is required to check the purity of plating salts.

Analytical methods are dependent on the purpose of the analysis, so that it is not possible to use a single method that will be satisfactory for all applications of a substance. A particular method is selected according to the relative importance of the factors: speed, cost and accuracy. Each of these factors assumes a different importance in each application. A method used to control the metal in a flash bath may not be satisfactory to control the metal in a bath used to produce heavy deposits. Likewise, methods used to control a bright plating bath may not be required for a bath used to produce deposits for resistance to corrosion.

Many times it is desirable to have more than one method for the same substance. Two different methods may be used to check the results of one against the other or an analyst may prefer a certain analysis. Of the many methods available, the following six methods are examples of various ways of determining copper in a copper cyanide bath:

1. Destroy the cyanides and plate the copper from an acid solution on a weighted platinum cathode.

2. Destroy the cyanides with acid and titrate with sodium thiosulfate.

3. Determine colorimetrically as the blue complex formed with ammonia or by the color produced from copper chloride in a strong hydrochloric acid solution.

4. Determine polarographically by the addition of the proper electrolyte.

5. Precipitate as copper hydroxide and determine the volume of precipitate with a centrifuge.

6. Estimate the copper from the titration of free cyanide and the cathode efficiency.

The selection of one of these methods depends on the accuracy desired and the equipment available. If a centrifuge is available for the determination of sulfate in a chromic acid bath, it may also be used for the determination of copper. If platinum electrodes are available for the electrolytic determination of copper, then these may be also used for the electrolytic determination of cadmium.

Volumetric Methods

The volumetric methods are very popular. They are relatively fast and usually quite accurate. They depend on the quantitative completion of a chemical reaction by addition of one solution to another. The volume of the solution that is added to the sample is accurately measured. The completion of the first reaction is indicated by a secondary reaction which proceeds with a change in color or with the formation of a precipitate. The accuracy of a volumetric method depends on:

1. The completeness of the primary reaction.
2. The sharpness of the end point indicated by the secondary reaction.
3. The concentration of the standard solution (usually expressed as normality). The chemists have determined the best reactions and the most suitable indicators for a particular method; but it is up to the analyst to prepare and test his standard solutions. For plating analyses, most standard solutions may be made by weighing enough chemical to make a solution of the desired strength. However, it should be realized that accurate solutions require standardization. It is not possible to make a solution of hydrochloric acid that will be $0.1000N \pm 0.0002N$ unless the solution is carefully standardized and adjusted. However, it is possible to make a silver nitrate solution of this accuracy without standardization, since silver nitrate is a pure salt and may be used as a standard for other solutions. As an example: Silver nitrate may be used to determine the value of a hydrochloric acid solution (by neutralizing a known volume of solution and titrating for chloride). Silver nitrate, however, is limited as a standard, since it must be freshly made. The difference between a fresh solution and an old one is very important. Most standard solutions decompose with age. Solutions must either be made up often, or they must be occasionally standardized. The best standard solutions are those that are stable, for example, a sodium chloride solution. The best methods for standardization are those that use

exactly the same steps as the analytical method. The best standard solution to standardize silver nitrate for the determination of chloride in a nickel bath would be a standard nickel bath. In other words, a nickel bath similar to that being analyzed with a known amount of chloride. As a matter of fact, the same nickel bath could be made to serve as a standard for sodium hydroxide solution for the determination of boric acid, and as a standard for sodium cyanide solution for the determination of nickel.

Gravimetric Methods

Gravimetric methods are not widely used for plating-bath control since they are slower than the other types of methods. The gravimetric methods consist of careful separation of a chemical compound from other substances in the bath. The compound is precipitated, washed, dried and weighed. These methods require a balance and an oven, depending on the drying step. The gravimetric methods are generally accurate and are often used for check work where a substance is determined by more than one method.

Electrolytic Methods

Electrolytic methods may come under the general classification of gravimetric methods since the metal is weighed. In the electrolytic methods the metal sought is deposited on a platinum cathode. It is then washed, dried and weighed. These methods are usually faster than the other gravimetric methods and are quite accurate. The high accuracy that can be obtained is indicated by the fact that the electric current is standardized in terms of plating silver from a silver nitrate solution.

Cathode-Efficiency Methods

The amount of metal in a bath can often be estimated by determination of the cathode efficiency. This may be done by plating at a known current density for a definite period of time. The amount of metal on the cathode may be determined by weighing or it may be determined by deplating at low current density and noting the time required. These methods can only be used if the effect of metal concentration, temperature and concentration of other chemicals on the cathode efficiency are known. The accuracy is only fair, but very rapid methods can be developed.

The Centrifuge

The centrifuge may be used for the rapid determination of an ingredient that can be precipitated.[1] Carbonate can be determined by the addition of barium chloride. Sulfate can be determined in a chromic acid bath. Free cyanide can be determined in a copper cyanide bath by the addition of copper sulfate solution. Many of the metals can be determined by centrifuging, if they can be precipitated as a sulfide, sulfate or other insoluble compound.

The pH Meter

The pH meter is valuable for the direct determination of pH and as a means of recognizing the end point where a titration is made to a definite pH.

The Colorimeter

Colorimetric methods are rapid and reasonably accurate where they can be applied. Specific methods must be worked out so that the color is measured at the proper wave length. Nickel can be determined colorimetrically in a few minutes. It is not necessary that the solution is colored if it will absorb light rays, as shown by a method used for the determination of nitrate in silver baths.[2]

Colorimetric methods are often valuable to determine impurities in plating baths with a view to their sensitivity to small amounts of metal.[3]

The Polarograph

The polarograph is an instrument that measures the voltage and amperage characteristics of a dropping mercury electrode.

It has been applied to the rapid simultaneous determination of copper and zinc in brass deposits.[4] It is obvious that this instrument can be used for the determination of the metal content of most of the plating baths. However, as with the colorimeter, much experimental work is required before it can be determined how the instrument will compete with available methods. It is suitable for the analysis of alloy baths and deposits.

The polarograph, like the colorimeter, can be used to determine small amounts of impurities in plating baths.[5]

Methods Still Needed

There are many ingredients that cannot be determined by known methods. Most of these are addition agents. These complex organic substances do not lend themselves to ready analysis; if they are colloidal, the analysis may be of no value. One may analyze for gelatin in an acid bath by determination of the total nitrogen content, but the analysis will not tell how the bath will operate. Only part of the gelatin participates actively in the process, the rest may only be a harmful residue. The only answer to this problem is to control the bath by regular addition, by observation of the plate or by plating-cell control.

Specific Analytical Methods

The following methods are specific for the analysis of the given substances in a plating bath. They should be used as a reference list of methods. Alternatives are given and new methods of analysis are suggested so that one may select one's own set of analytical procedures. It is well for the plater to do this since each application may call for a different method, depending on the equipment available and the accuracy required. The list indicates that many methods can be developed that are not specific plating methods. New methods may be developed from standard analytical texts or from a knowledge of general methods such as the application of a pH meter to analytical work.

Determination of Acids

Boric Acid in a Nickel Bath

Pipette a 5 ml. sample into a 250 ml. Erlenmeyer flask. Add 25 ml. glycerin. Add 0.5 ml. bromcresol purple indicator solution. Titrate with N/10 sodium hydroxide until the solution changes to dark green and then to a blue purple.

$$\text{ml. NaOH} \times 0.166 = \text{oz./gal. } H_3BO_3$$

It is well to standardize the sodium hydroxide with a nickel bath containing a known amount of boric acid. The same size of sample and exactly the same procedure should be used as in the analysis. This will enable the analyst to recognize the proper color change and all errors in the method will be alike during the analysis and the standardization, thus canceling the errors.

Undersaturation Method

Add 3 to 6 g. boric acid to 250 ml. of the bath. Add a few drops of wetting agent (e.g., Tergitol). Agitate for 1 hour. Filter on a small Buchner funnel. Wash with 10 ml. acetone and allow to dry. Pull dry for 15 minutes. The residue contains 20 per cent liquid.

In some plating baths, such as the fluoborate bath, boric acid is very difficult to determine. This method, which may be called a measurement of undersaturation may be used in such a case. If it is used, it is best to redefine the limits for boric acid in terms of undersaturation. For instance, if the limits call for 25 to 35 g./l. and a saturated bath would contain 45 g./l., then the undersaturation limits are 10 to 20 g./l. The residue weighed is a direct measure of undersaturation where the residue holds 20% liquid. A correction should be made for dry weight or the residue should be dried.

Chromic Acid

Pipette 10 ml. of the sample into a 500 ml. volumetric flask. Dilute to the mark and mix thoroughly. Pipette 10 ml. from the volumetric flask into a 250 ml. Erlenmeyer flask. Add 100 ml. distilled water. Add 2 g. ammonium bifluoride. Add 15 ml. concentrated hydrochloric acid solution and 10 ml. of a solution containing: 100 g./l. potassium iodide and 1 g./l. potassium hydroxide. Titrate with N/10 sodium thiosulfate until the brown color fades to straw. Add 5 ml. starch solution and titrate until the blue color disappears.

$$\text{ml. } Na_2S_2O_3 \times 2.234 = \text{oz./gal. } Cr.O_3$$

Fluoboric Acid

Fluoboric acid may be determined by conductivity if correction is made for the metal content of the bath.

The correction for the metal content may be made by a gravity measurement.

Free Fluoboric Acid in a Lead Fluoborate Bath

Pipette a 10 ml. sample into a 400 ml. beaker. Dilute to 250 ml. with distilled water. Titrate with N/1 sodium hydroxide until the first permanent white cloudiness appears.

In most analytical procedures, recommended volumes are only approximate, but in this procedure the recommended volume must be exactly the same for each determination.

$$\text{ml. NaOH} \times 1.18 = \text{oz./gal. } HBF_4$$

Free Acid

Pipette a 25 ml. sample into a 250 ml. Erlenmeyer flask. Add 125 ml. distilled water and 5 drops of methyl orange indicator. Titrate with N/1 sodium hydroxide.

$$\text{ml. NaOH} \times 0.196 = \text{oz./gal. HCl}$$
$$\text{ml. NaOH} \times 0.263 = \text{oz./gal. } H_2SO_4$$

This is a general method for free acid in the presence of any metal that does not precipitate before the methyl orange end point is reached.

Hydrochloric Acid and Iron

Hydrochloric acid can be determined in the presence of iron by the conductivity method, but a correction must be made for the iron. This correction can be made by a gravity measurement. By properly setting up original standards, acid and iron can thus be run by the measurement of conductivity and gravity.

Sulfuric Acid in Acid Copper

Pipette a 10 ml. sample into a 250 ml. beaker. Add 100 ml. distilled water and 5 drops of methyl orange indicator. Titrate with N/1 sodium hydroxide until the color of the solution changes from violet to pale green.

$$\text{ml. NaOH} \times 0.657 = \text{oz./gal. } H_2SO_4$$

Ammonia

Cyanide Solutions

Pipette a 20 ml. sample into a 1 l. distilling flask. Place 75 ml. 10 per cent sodium hydroxide solution in the dropping funnel. Pipette 25 ml. of N/10 hydrochloric acid into the receiver and add 1 drop of methyl red indicator. Slowly run the sodium hydroxide solution into the distilling flask, leaving a small amount in the funnel to act as a seal. Distill off the ammonia and watch that the color of the indicator does not change. Back-titrate the excess hydrochloric acid with N/10 sodium hydroxide. Run a blank to determine the exact milliliters of N/10 sodium hydroxide required to titrate 10 ml. of N/10 hydrochloric acid.

$$(\text{ml. blank—ml. titration}) \times 0.0234 = \text{oz./gal. } NH_4OH$$

Brass Bath

Monaweck[6] reported that the precipitates of $NaCU(CN)_2$ and CuCN decomposed during distillation to convert CN to NH_3. To eliminate these precipitates, he suggests the following method: Pipette a 10 cc. sample of the brass solution into an Erlenmeyer flask. Add 25 cc. distilled water. Add 25 cc. of cold 1 to 4 H_2SO_4. Filter off precipitate. Wash precipitate with 50 cc. water. Evaporate filtrate until SO_3 fumes begin to appear. Cool and dilute to 250 cc. Add a few drops of phenolphthalein. Add $Ca(OH)_2$ solid, until the slurry turns pink. Distill for ½ hour in a Kjeldahl apparatus, collecting the NH_3 distillate in standard acid. Titrate the excess acid.

$$(\text{ml. blank—ml. titration}) \times 0.0468 = \text{oz./gal. } NH_4OH$$

Cadmium

Cyanide Solutions—Alcohol Method

Pipette 5 ml. sample into a 25 ml. volumetric flask. Make up to the mark with methyl alcohol. Shake and allow the crystals to settle. Filter dry into a beaker. Pipette 10 ml. into a 100 ml. beaker. Add 10 ml. 1 to 2 hydrochloric acid and 40 ml. water. Titrate with N/10 potassium ferrocyanide.

$$\text{ml. } K_4Fe(CN)_6 \times 0.752 = \text{oz./gal. Cd.}$$

Cyanide Solutions—Electrolytic Method

Pipette a 5 ml. sample into a 400 ml. electrolytic beaker. Add 75 ml. distilled water. Electrolyze with platinum electrodes at 1.5 amperes for 2 hours, using a weighed cathode. Dry and weigh the cathode.

$$\text{Cd deposited} \times 26.8 = \text{oz./gal.}$$

Cyanide Solutions—Volumetric Method

Pipette a 5 ml. sample into a 400 ml. beaker. Add 50 ml. distilled water and 10 ml. of 15 per cent sodium sulfide. Warm the solution and filter off the cadmium sulfide precipitate. Wash the precipitate thoroughly with warm distilled water and discard the filtrate. Transfer the paper and precipitate into the original beaker. Add 25 ml. concentrated hydrochloric acid and 25 ml. distilled water. Heat to boiling to dissolve the precipitate. Dilute to 200 ml. with distilled water and add ammonium hydroxide until it turns the red litmus paper blue. Add concentrated hydrochloric acid until the litmus paper turns red again and then add 3 ml. in excess. Heat nearly to boiling and titrate with N/10 potassium ferrocyanide using uranium acetate as an outside indicator.

The presence of zinc will cause an error in this method.

$$\text{ml. } K_4Fe(CN)_6 \times 0.300 = \text{oz./gal. Cd.}$$

Carbonate

Cyanide Solutions

Pipette a 10 ml. sample into a 150 ml. beaker. Add 1 ml. concentrated ammonium hydroxide and 100 ml. 10 per cent barium chloride solution. Heat to boiling and boil gently for 10 minutes. Filter and wash the precipitate with hot water. Add a few drops of barium chloride solution to the filtrate to determine if the precipitation was complete. Transfer the filter paper and precipitate into the original beaker. Add 50 ml. distilled water and a few drops of methyl orange indicator. Titrate with N/1 hydrochloric acid to a permanent pink color.

$$\text{ml. HCl} \times 0.71 = \text{oz./gal. } Na_2CO_3$$

The accuracy of this method is only fair, which is all that is usually required. Greater accuracy may be obtained by distilling the carbon dioxide with sulfuric acid, collecting it by precipitation as it is bubbled through a barium hydroxide solution and then treating the precipitate as above. For greater detail on carbonate analyses consult the analytical books.

Cyanide Baths[7]*—Centrifuge Method*

Pipette a 5 ml. sample into a 100 ml. Goetz centrifuge tube. Add 4 drops of wetting agent and 45 ml. distilled water. Shake the tube so that its contents are thoroughly mixed. Place on a water bath above 200°F. for 5 minutes. Add 50 ml. barium chloride solution (25 g./l.) also above 200°F. Shake and then place on a hot water-bath for 10 minutes. Remove from the bath, shake and then spin in centrifuge at 2,000 r.p.m. for 5 minutes. Read the height of the precipitate in the tube and derive the concentration of sodium carbonate in ounces per gallon from a prepared standard curve.

Chloride

Pipette a 5 ml. sample into a 250 ml. Erlenmeyer flask. Add 50 ml. distilled water and 1 ml. 10 per cent potassium chromate solution. Titrate with standard N/10 silver nitrate solution until a reddish color appears.

$$\text{ml. } AgNO_3 \times 0.095 = \text{oz./gal. Cl}$$

Chromium

Trivalent Chromium in a Chromic Acid Bath

Pipette a 10 ml. sample into a 500 ml. volumetric flask. Dilute to the mark and mix thoroughly. Pipette 10 ml. from the volumetric flask into a 250 ml. Erlenmeyer flask. Add 100 ml. distilled water. Add about 0.2 g. sodium peroxide. Boil for 20 to 30 minutes. Dilute to 100 to 125 ml. and cool. Add 2 g. ammonium bifluoride and proceed as for the analysis of chromic acid.

$$\text{ml. } Na_2S_2O_3 \text{ used for } Cr^{+++} - \text{ml. } Na_2S_2O_3 \text{ used for } CrO_3 \times 1.16 = \text{oz./gal. } Cr^{+++}$$

Copper

Electrolysis Method for Cyanide Solutions

Pipette a 20 ml. sample into a 200 ml. electrolytic beaker. Add 5 ml. concentrated nitric acid and 5 ml. concentrated sulfuric acid. Boil until dense white fumes of sulfur trioxide are evolved. Cool. Slowly add 100 ml. distilled water. Weigh a platinum-gauze cathode and set up the platinum cathode and a platinum anode for electrolysis. Lower the electrodes into the beaker until the solution nearly covers them and start agitation (agitate with a stirrer or with air). Deposit the copper at 4.0 amperes until the blue color disappears and then continue the electrolysis for an additional 15 minutes at 2.0 amperes. Add water so that the level of the solution is raised and an unplated portion of the cathode is exposed to the current. If no further copper deposits, the electrolysis is complete. Without interrupting the current, stop the agitation and slowly raise the electrodes while washing the cathode down with a continuous stream of water to prevent resolution of the copper. Rinse the cathode in alcohol, then in ether and dry at 50°C. Weigh. From the gain in weight determine the copper content.

$$\text{g. Cu} \times 6.7 = \text{oz./gal. Cu}$$

Volumetric Method for Acid Solutions

Pipette a 2 ml. sample into a 250 ml. beaker. Add 25 ml. distilled water. Add concentrated ammonium hydroxide until the solution turns dark blue. Boil for 15 minutes. Add 10 ml. 30 per cent acetic acid solution. Cool to room temperature. Add 25 ml. 20 per cent potassium iodide solution. Swirl the contents to mix. Titrate with N/10 sodium thiosulfate using starch as an indicator.

$$\text{ml. } Na_2S_2O_3 \times 0.426 = \text{oz./gal. Cu.}$$

Volumetric Method for Cyanide Solutions

Pipette a 10 ml. sample into a 250 ml. Erlenmeyer flask. Add 5 ml. concentrated sulfuric acid and 0.5 ml. concentrated nitric acid. Boil until dense white sulfur trioxide fumes are evolved. Cool. Add 100 ml. distilled water. Add concentrated ammonium hydroxide until the solution turns dark blue. Boil for 15 minutes. Add 10 ml. 30 per cent acetic acid solution. Cool to room temperature. Add 25 ml. 20 per cent potassium iodide solution. Swirl the contents to mix. Titrate with N/10 sodium thiosulfate solution until the brown color fades to yellow. Add 5 ml. 1 per cent starch solution and continue titrating until the blue color disappears and does not reappear for at least 1 minute.

$$\text{ml. } Na_2S_2O_3 \times 0.0852 = \text{oz./gal. Cu}$$

Cyanide

Free Cyanide in Copper Cyanide

Pipette a 5 ml. sample into a 150 ml. beaker. Add 100 ml. distilled water. Add a few drops of 10 per cent potassium iodide solution. Titrate with N/10 silver nitrate to a slight turbidity.

$$\text{ml. } AgNO_3 \times 0.261 = \text{oz./gal. free NaCN}$$

This method is reliable for control of a Rochelle copper bath, but it must be modified to control a high-efficiency copper cyanide bath.

Free Cyanide in a High-Efficiency Copper Bath

Pipette 10 ml. of the bath into a dry flask. Add 3 g. Rochelle salt and 1 g. potassium iodide. Agitate until the salts are dissolved. Titrate with N/10 silver nitrate.

$$\text{ml. } AgNO_3 \times 0.130 = \text{oz./gal. free NaCN}$$

This method will give approximately the same result as the cold titration for a high-efficiency sodium bath. For the potassium bath, it is believed to be more reliable than the cold titration.

Free Cyanide in High-Efficiency Copper Bath

Place 10 ml. of 10 per cent potassium iodide solution and 40 g. cracked ice into a 250 ml. Erlenmeyer flask. Pipette 10 ml. plating solution into the flask and stir until a temperature of 3 to 5°C. is reached. Titrate slowly at this temperature with N/10 silver nitrate until a distinct opalescence appears which is permanent for 1 minute.

$$\text{ml. } AgNO_3 \times 0.131 = \text{oz./gal. free NaCN}$$

Total Sodium Cyanide in Cadmium Cyanide

Pipette 2 ml. sample into a 250 ml. Erlenmeyer flask. Add 100 ml. distilled water, 5 ml. 100 per cent potassium iodide and 15 ml. concentrated ammonium hydroxide. Titrate with N/10 silver nitrate to a faint yellow turbidity.

$$\text{ml. } AgNO_3 \times 0.65 = \text{oz./gal. NaCN}$$

Total Sodium Cyanide in Cadmium Cyanide

Pipette a 1 ml. sample into a 250 ml. Erlenmeyer flask. Add 100 ml. distilled water. Add 5 g. of a dry powdered mixture of 95 per cent by weight sodium bicarbonate and 5 per cent by weight potassium chromate. Dissolve the salts. Titrate with N/10 silver nitrate containing 10 ml. per liter concentrated nitric acid to the first permanent brick-red color.

$$\text{ml. } AgNO_3 \times 1.30 = \text{oz./gal. NaCN}$$

If iron or ferrocyanide is present, it will introduce an error in this method.

Free Cyanide in Silver and Gold Baths

The general method for total sodium cyanide may be used for the determination of free cyanide in gold and silver baths.

Total Sodium Cyanide

Pipette a 10 ml. sample into a 250 ml. Erlenmeyer flask. Add 20 ml. 10 per cent potassium iodide solution and 100 ml. distilled water. Titrate with N/10 silver nitrate to the first turbidity.

$$\text{ml. } AgNO_3 \times 0.130 = \text{oz./gal. free NaCN}$$

This is the general method for determining sodium cyanide. If sodium cyanide alone is present it will measure the total sodium cyanide present. In a silver bath, it will only measure the free cyanide. The method may be altered for each bath, depending on the metal present.

Total Cyanide in Zinc Cyanide Solutions

Pipette a 5 ml. sample into a 250 ml. Erlenmeyer flask. Add 100 ml. distilled water, 5 ml. 20 per cent sodium hydroxide solution, and 5 ml. 10 per cent potassium iodide solution. Titrate with N/10 silver nitrate to a faint turbidity.

$$\text{ml. } AgNO_3 \times 0.261 = \text{oz./gal. total NaCN}$$

Gelatin

Take a 25 ml. sample. Add 15 ml. sulfuric acid, 4 g. potassium hydrosulfate and 0.1 g. selenium. Digest for 30 minutes. Dilute to 300 ml. Add 40 ml. 40 per cent sodium hydroxide. Distill into 10 ml. N/10 hydrochloric acid.

Titrate the remaining hydrochloric acid with N/10 sodium hydroxide.

Run a blank titration on 10 ml. of N/10 hydrochloric acid.

$$(\text{ml. blank} - \text{ml. titration}) \times 0.0502 = \text{oz./gal. gelatin}$$

Gold

Analysis of Gold and Gold-Alloy Solutions

1. Pipette 25 ml. gold plating solution (or the volume which contains 20 to 50 mg. gold) into a 400 ml. beaker. This size sample is correct for a 3 to 5 pennyweight per gallon bath.

2. Under a fume hood, add 25 ml. concentrated hydrochloric acid, stir and bring to a boil. Add 5 ml. concentrated nitric acid and boil for 5 minutes.

3. Remove from flame and add slowly, with agitation, 25 ml. of a 5 per cent sodium hypochlorite solution. Then add 35 ml. water. Mix thoroughly.

4. Boil gently for 10 minutes.

5. Allow to cool and add, with stirring, 20 ml. 20 per cent sodium hydroxide solution. From a pipette or burette, add a saturated solution of sodium bicarbonate in portions of 5 to 7 ml. After each addition, stir vigorously and test the acidity by touching a piece of red litmus paper with the end of the stirring rod. The drop on the stirring rod should be as small as possible. A faint trace of blue in the litmus paper is taken as the end point for the addition of bicarbonate.

If, after the addition of sodium hydroxide, the sample is alkaline, add 1 ml. dilute hydrochloric acid at a time until the sample is acid to litmus. Then add bicarbonate as described above.

6. Add 20 ml. 30 per cent potassium iodide solution all at once with agitation. Allow to stand for 2 minutes.

7. Add 2 g. solid sodium bicarbonate and titrate with a standard N/100 arsenious oxide solution.

8. When the yellow-red iodide color becomes a pale straw yellow, add 2 ml. of a 1 per cent starch solution and continue the titration to the disappearance of the starch-iodine color.

In the presence of blue-colored basic compounds, i.e., copper, it is best to add the starch solution at the beginning.

Cyanide Solution

Pipette a 10 ml. sample into a 250 ml. beaker. Add 15 ml. concentrated hydrochloric acid and evaporate to a syrupy consistency over a water- or steam-bath. Add 150 ml. of distilled water. Add 25 ml. 20 per cent potassium iodide solution and 2 ml. 1 per cent starch solution. Titrate with N/100 sodium thiosulfate to complete disappearance of the blue color. Back titrate with N/100 iodine solution to reappearance of the blue color.

Reagents of N/100 normality are not stable enough to use without standardization. Therefore, these reagents must be standardized with a known amount of gold.

Electrolytic Method: Pipette a 5 ml. sample into a 250 ml. electrolytic beaker. Add 200 ml. distilled water. Add 5 g.

sodium hydroxide and 5 g. sodium cyanide and dissolve. Electrolyze at 4 to 6 volts for about 45 minutes on a weighed platinum cathode. Rinse with distilled water, dry in an oven, cool and weigh.

$$\text{g. Au} \times 200 = \text{g./l. Au}$$
$$\text{g. Au} \times 26.8 = \text{oz./gal. Au}$$

Indium

Cyanide Bath

Pipette a 10 ml. sample into a 250 ml. beaker. Add 10 ml. concentrated hydrochloric acid. Add 25 ml. distilled water and boil to expel the hydrogen cyanide. Add 5 drops of methyl-orange indicator. Add concentrated ammonium hydroxide slowly until the solution turns yellow. Bring to a boil. Filter and wash. Ignite in a good oxidizing atmosphere at about 800°C. Weigh as In_2O_3.

$$\text{oz. } In_2O_3 \times 11.1 = \text{oz./gal. In}$$

Iron

Chloride Bath

Pipette a 5 ml. sample into a 250 ml. Erlenmeyer flask. Add 125 ml. distilled water and 15 ml. of sulfuric-phosphoric mixture consisting of 150 ml. concentrated sulfuric acid and 150 ml. concentrated phosphoric acid diluted to 1 l. Add a few drops of diphenylamine solution consisting of 1 g. diphenylamine dissolved in 100 ml. 60 per cent acetic acid. Titrate with N/2 potassium dichromate until the solution changes from green to violet.

$$\text{ml. } K_2Cr_2O_7 \times 0.748 = \text{oz./gal. Fe}$$

Sulfate Bath

Pipette a 5 ml. sample into a 250 ml. Erlenmeyer flask. Add 100 ml. distilled water and 5 ml. sulfuric acid solution consisting of 1 part sulfuric acid to 1 part water. Titrate with N/2 potassium permanganate to a permanent pink color.

$$\text{ml. } KM_nO_4 \times 0.748 = \text{oz./gal. Fe}$$

Lead

Gravimetric Method

Pipette a 10 ml. sample into a 400 ml. beaker. Add 100 ml. distilled water. Add 10 per cent sulfuric acid until precipitation is complete. Heat almost to boiling and keep hot for 15 minutes. Cool. Filter and wash with cold 1 per cent sulfuric acid. Ignite at 800°C. Weigh as $PbSO_4$.

$$\text{g. } PbSO_4 \times 9.1 = \text{oz./gal. Pb}$$

Volumetric Method

Pipette a 5 ml. sample into a 400 ml. beaker. Add 10 ml. concentrated nitric acid and 15 ml. concentrated sulfuric acid. Evaporate to dense white fumes. Cool. Add 150 ml. distilled water and 10 g. tartaric acid. Heat to boiling. Cool. Filter off the lead sulfate and wash with cold water. Transfer the paper to the original beaker and add 30 ml. saturated ammonium acetate solution. Add 150 ml. distilled water. Heat to boiling and titrate the hot solution with standard ammonium molybdate solution using tannic acid as an external indicator. The end point is the first faint yellow color.

$$(NH_4)_2MoO_4 \times 0.134 = \text{oz./gal. Pb}$$

Nickel

Pipette a 5 ml. sample into a 250 ml. Erlenmeyer flask. Add 50 ml. distilled water and 0.5 ml. 10 per cent potassium iodide solution. Add concentrated ammonium hydroxide until the solution turns a clear blue. Add 0.5 ml. N/10 silver nitrate. Titrate with N/2 sodium cyanide until the precipitate which first forms just redissolves.

$$\text{ml. NaCN} \times 0.393 = \text{oz./gal. Ni}$$

Resorcinol

Dilute a 10 ml. sample to 50 ml. Add 2 g. 20-mesh zinc and allow to stand for 10 minutes. Add 50 ml. water and 5 ml. hydrochloric acid and allow to stand 30 minutes longer. Filter and wash. Add potassium bromide and standard potassium bromate solution. Allow to stand for 15 minutes. Add potassium iodide. Allow to stand 15 minutes and titrate with N/10 sodium thiosulfate. Run a blank without the sample and the zinc, but with the timing the same as with the sample, to determine the value of 50 ml. potassium bromate.

(ml. blank — ml. titration) $\times$ 0.0245 = oz./gal. resorcinol

Potassium Bromate Solution	3 g./l.
Potassium Bromide Solution	250 g./l.
Potassium Iodide Solution	25 per cent

For bromination methods of other organic compounds, see Kolthoff and Furman, Vol. II, p. 476.

Rochelle Salt

Cyanide Solutions

Pipette a 5 ml. sample into a 250 ml. volumetric flask. Add 100 ml. distilled water and a particle of solid phenolphthalein. Add 1 per cent sulfuric acid solution, drop by drop, until the pink color just disappears. Add exactly 5 ml. nitrobenzene. Add 10 per cent silver nitrate solution, drop by drop, until the color of the precipitate changes from white to blue-gray. Shake ½ minute to coagulate the precipitate. Allow to settle and add 1 drop silver nitrate to determine if the precipitation is complete. Bring the solution up to the mark, add exactly 5 ml. distilled water above the mark and shake thoroughly. Allow to stand until the precipitate has settled. Pipette 50 ml. into a 500 ml. Erlenmeyer flask. Add 5 ml. 20 per cent sulfuric acid, 5 g. manganese sulfate and 100 ml. distilled water. Heat to about 70°C. Add slowly and with agitation exactly 20 ml. N/10 potassium permanganate solution. Allow the hot solution to stand for 5 minutes. Cool to room temperature under running water. Add 2 g. solid potassium iodide and 2 ml. 1 per cent starch solution. Shake the flask for a few seconds. Titrate with N/10 sodium thiosulfate until the blue color disappears and does not reappear for at least 1 minute.

The potassium permanganate and sodium thiosulfate should be standardized for this method.

$$6.28\ (20 \times \text{N of } KMnO_4 - \text{ml. } Na_2S_2O_3 \times \text{N of } Na_2S_2O_3 = \text{oz./gal. } KNaC_4H_4O_6.4H_2O$$

Silver

Pipette a 5 ml. sample into a 250 ml. beaker. Add 100 ml. distilled water and 10 ml. of 15 per cent sodium sulfide solution. Heat to boiling. Filter and wash the precipitate with hot water. Transfer the paper and precipitate to the original beaker and add 50 ml. 1 to 2 nitric acid. Boil to dissolve the precipitate. Add 10 ml. 10 per cent ferric ammonium sulfate. Titrate with N/10 potassium thiocyanate solution to the appearance of a pink coloration.

$$\text{ml. KCNS} \times 0.372 = \text{oz./gal. AgCN}$$

Electrometric Determination

This method illustrates how a cell may be set up to titrate to a definite potential. The method is similar to titration to a definite pH with a pH meter except that a system has to be used that will measure the potential of the metallic ion rather than that of the hydrogen ion.

An apparatus was developed so that both free cyanide and silver could be determined from one titration.

Cyanide Strike Bath

Silver can be determined in a silver-strike bath by plating and deplating. This is a rapid method of determining the cathode efficiency which is a measure of the silver concentration. By dilution of a silver bath with a sodium cyanide solution, this method can be used for the determination of silver in a standard silver cyanide bath.

Cyanide Bath

Pipette 10 ml. bath into a 400 ml. beaker. Dilute to 100 ml. Add nitric acid, drop by drop, until the color of phenolphthalein is just discharged. Boil to coagulate the precipitate. Filter in a weighed Gooch or fritted-glass crucible. Wash. Dry at 105°C. for ½ hour and weigh as silver cyanide.

$$\text{g. Ag. CN} \times 10.8 = \text{oz./gal. Ag}$$

Sodium Acetate

Pipette a 25 ml. sample into a 500 ml. distilling flask. Place a few glass beads on the bottom of the flask. Fit the flask with a two-hole rubber stopper, one hole containing a thermometer and the other a dropping funnel. Connect the side arm of the flask to a condenser. Add 175 ml. distilled water and 25 ml. concentrated sulfuric acid through the dropping funnel. Distill into a 500 ml. Erlenmeyer flask containing 50 ml. distilled water. Continue the distillation until the temperature reaches about 127°C. Add more water and continue the distillation until 250 to 300 ml. water has been distilled over. Add 0.5 ml. sulfo-orange indicator and titrate with N/2 sodium hydroxide to a yellow end point.

$$\text{ml. NaOH} \times 0.220 = \text{oz./gal. } Na_2C_2H_3O_2$$

Sodium Hydroxide

Pipette a 5 ml. sample into a 250 ml. Erlenmeyer flask. Add 25 ml. distilled water and 0.5 ml. sulfo-orange indicator. Titrate with N/10 hydrochloric acid to a yellow end point.

$$\text{ml. HCl} \times 0.107 = \text{oz./gal. NaOH}$$

Cadmium Cyanide Bath for Sodium Hydroxide

Pipette a 5 ml. sample into a 250 ml. Erlenmeyer Flask. Add 50 ml. distilled water and 10 ml. 10 per cent sodium cyanide.

Add 0.5 ml. sulfo-orange indicator. Titrate with standard N/1 hydrochloric acid until the color changes from orange to yellow.

$$\text{ml. HCl} \times 0.0107 = \text{oz./gal. NaOH}$$

Sodium Cyanide

Sodium hydroxide can be controlled by conductivity in the presence of sodium cyanide, if the sodium cyanide is determined by titration and a correction for the effect of sodium cyanide on the conductivity is determined.

Sodium Thiocyanate

Pipette a 10 ml. sample into a 250 ml. Erlenmeyer flask. Add 100 ml. distilled water. Add sulfuric acid, drop by drop, until the solution just turns blue litmus red. Add 0.5 ml. saturated ferric ammonium sulfate solution. Titrate with standard N/10 silver nitrate solution until the red color disappears.

$$\text{ml. } AgNO_3 \times 0.109 = \text{NaCNS}$$

Sulfate

Chromic Acid Baths

Pipette a 10 ml. sample into a 250 ml. beaker. Add 75 ml. of a reducing mixture consisting of 15 parts by volume isopropanol, 7 parts concentrated hydrochloric acid and 25 parts glacial acetic acid. Dilute with hot water to between 125 and 150 ml. and lct stand for at least 1 hour at 60 to 70°C. Filter into a 250 ml. beaker and wash the green coloration from the paper with hot water. Heat the filtrate to boiling and add, drop by drop, 10 ml. 10 per cent barium chloride solution. Allow to settle for 2 to 4 hours at 60 to 70°C. Filter and wash thoroughly. Ignite the paper and precipitate in a weighed porcelain crucible at bright red heat for about 15 minutes. Cool and weigh.

$$\text{g. BaSO} \times 5.64 = \text{oz./gal. } H_2SO_4$$

Chromic Acid Ratio

The $CrO_3 : SO_4$ ratio of a chromic acid bath can be determined directly from a plating-range test, if plating range standards are available. This method has the advantage that the bath can be adjusted to the widest plating range by test rather than relying on chemical analysis for SO_4 to produce the widest plating range.

Tin

Alkaline Tin Bath

Tin can be determined in an alkaline tin bath by using a rapid plating and deplating method for the measurement of cathode efficiency. If such a method is used, the temperature and caustic concentration effects will have to be taken into account.

Stannous Tin in an Acid Tin Bath

Pipette a 10 ml. sample into a 500 ml. Erlenmeyer flask. Add 150 ml. distilled water, 20 ml. concentrated hydrochloric acid and 2 ml. starch solution. Titrate with N/10 iodine solution.

$$\text{ml. I} \times 0.0795 = \text{oz./gal. Sn}$$

Total Tin in an Acid Bath

Pipette a 10 ml. sample into a 500 ml. Erlenmeyer flask. Add 150 ml. distilled water, 10 ml. concentrated hydrochloric acid and 1 g. powdered iron. Attach an outlet tube and boil for 30 minutes to dissolve the iron and reduce the tin. Place the flask in a cooling bath with the outlet tube submerged in saturated sodium bicarbonate solution. When cool, remove the outlet tube, add starch and titrate with N/10 iodine solution.

$$\text{ml. I} \times 0.0795 = \text{oz./gal. Sn}$$

Zinc

Cyanide Solution No. 1

Pipette a 5 ml. sample into a 25 ml. volumetric flask. Make up to the mark with methyl alcohol. Shake and allow the crystals to settle. Filter dry into a beaker. Pipette 10 ml. filtrate into a 100 ml. beaker. Add 10 ml. 1 to 2 hydrochloric acid and 40 ml. water. Add 1 ml. 10 per cent sodium sulfide. Titrate with N/10 potassium ferrocyanide.

$$\text{ml. } K_4Fe(CN)_6 \times 0.438 = \text{oz./gal. Zn}$$

Cyanide Solution No. 2

Pipette a 5 ml. sample into a 250 ml. beaker. Add 25 ml. distilled water and 10 ml. 15 per cent sodium sulfide solution. Heat to boiling and filter off the precipitate. Wash with hot water and discard the filtrate. Transfer the filter paper and precipitate into the original beaker. Add 10 ml. concentrated hydrochloric acid and 1 to 2 g. sodium sulfide crystals. Boil to expel the hydrogen sulfide. Dilute to 150 ml. with distilled water and heat to boiling. Titrate with N/10 potassium ferrocyanide using 5 per cent uranium nitrate as an outside indicator. The end point is reached when the addition of 1 drop of the solution being titrated to 2 drops of uranium nitrate on the spot plate produces a reddish-brown precipitate.

$$\text{ml. } K_4Fe(CN)_6 \times 0.175 = \text{oz./gal. Zn for N/10}$$

References

1. George Jernstedt, *Trans. Electrochem. Soc.,* **82,** 135 (1942).
2. A. Dolance and P. W. Healy, *Ind. Eng. Chem. Anal. Ed.,* **17,** 718 (1945).
3. E. A. Parker, *Monthly Rev., Am. Electroplaters' Soc.,* **34,** 33 (1947).
4. H. E. Zentler-Gordon and E. R. Roberts, *Trans. Electrochem. Soc.,* Preprint 90.92 (1946).
5. V. S. Baleya, N. I. Puchenkina and I. A. Korshunov, *Zavoskaya Lab.,* **11,** 644 (1946).

6. Monaweck, *Trans. Electrochem. Soc.*, **82**, 59 (1942).
7. G. Jernstedt, *Trans. Electrochem. Soc.*, **82**, 137 (1942).

BIBLIOGRAPHY

Modern Electroplating, The Electrochemical Society.

Plating and Finishing Guidebook, Finishing Publications, Inc.

Simple Methods of Analyzing Plating Solutions, Hanson-Van Winkle-Munning Co.

Year Book, Newark Branch, American Electroplaters' Society.

Scott, *Standard Methods of Chemical Analysis*, D. Van Nostrand Co.

Appendix

Instruments Used for Electroplating *

Instrument	*Use*
Magna-Gage	Thickness
Drop Testing Equipment	Thickness
Conductivity Meter	Conductivity and Solution Control
pH Meter	Solution Control
Hull Cell	Plating Range
Bent Cathode	Plating Range
Colorimeter	Solution Control
Polarograph	Solution Control
Salt Spray Cabinet	Specification Plating
Humidity Cabinet	Corrosion Resistance
Centrifuge	Rapid Analytical Methods

Plating Tests †

Electroplates are tested for specification and experimental purposes. The following table indicates the common tests.

Thickness Methods
- Magnetic
- Chemical Solution
- Anodic Stripping
- Chemical Solution Followed by Chemical Analysis

Porosity Methods
- Ferroxyl Test
- Black-Light Test

---

* For details on the construction and use of these instruments *see:*

Modern Electroplating, The Electrochemical Society, New York.

Plating and Finishing Guidebook, Metal Industry Publishing Co., New York.

Research Project Reports, American Electroplaters' Society, Jenkintown, Pennsylvania.

† Detailed description of these tests can be found in the publications listed in the footnote on the preceding page.

Adhesion Methods
- Bending
- Pounding
- Chiseling
- Scratching
- Burnishing
- Shot Blasting

Corrosion Methods
- Salt Spray
- Special Chemical Solutions

Conversion Factors

From	*To*	*Multiply by*
g./l.	oz./gal.	0.134
oz./gal.	g./l.	7.5
amp./sq.ft.	amp./sq.dm.	0.108
amp./sq.dm.	amp./sq.ft.	9.29
g./l.	oz./cu.ft.	1.00
in.	cm.	2.54
cm.	in.	0.394
gal.	l.	3.78
gal.	cu.ft.	0.134
l.	gal.	0.264
l.	cu.ft.	0.0353
cu.ft.	l.	23.3
cu.cm.	cu.in.	0.0610
cu.in.	cu.cm.	16.4
sq.cm.	sq.in.	0.155
sq.in.	sq.cm.	6.45

Strike Baths

Copper Strike

Copper Cyanide	4 oz./gal.
Sodium Cyanide	6 oz./gal.
Temperature	Room
Voltage	6
Time	1 minute

Silver Strike

Silver Cyanide	¼ oz./gal.
Sodium Cyanide	5 oz./gal.
Temperature	Room

Voltage	6
Time	1 minute

Nickel Strike

Nickel Sulfate	10 oz./gal.
Sodium Sulfate	15 oz./gal.
Boric Acid	4 oz./gal.
Ammonium Chloride	3 oz./gal.
Temperature	Room
Current Density	25 amp./sq.ft.
pH	5.8 to 6.2

Etching and Pickling Solutions

Aluminum

Nitric acid and hydrofluoric acid used separately or in combination.

Zinc

Hydrochloric acid 2 to 5 per cent, cold.

Silver-base and Copper-base

Nitric acid or anodic treatment in 5 per cent sodium cyanide.

Steel

Chemical pickling in 5 to 15 per cent hot hydrochloric or sulfuric acid or anodic electrolytic pickling in hydrochloric or sulfuric acid.

Alloy Steels

Alloy steels require special treatment. Anodic treatment in a chromic acid bath is usually successful prior to chromium plating.

Alkaline Cleaning Solutions

There are many proprietary alkaline cleaners on the market, but simple solutions can be made up that will do an efficient cleaning job. The primary chemicals required to make a cleaner for any type of metal are: sodium hydroxide, trisodium phosphate, sodium carbonate and sodium metasilicate. In addition, soaps and wetting agents are sometimes used and sodium hexametaphosphate may be added for free rinsing. Soap is usually added for soak cleaning and wetting agents to form a foam to trap spray for electrolytic cleaning.

High-pH cleaners are used for steel and copper alloys and lower-pH cleaners for active metals such as zinc and aluminum. Typical cleaner formulas are as follows:

Aluminum (Cleaning and Etching)

Sodium Carbonate	1 oz./gal.
Trisodium Phosphate	1 oz./gal.
Temperature	160°F.
Time	1 minute soak

Zinc-Base

Sodium Carbonate	4 oz./gal.
Trisodium Phosphate	2 oz./gal.
Sodium Hydroxide	1 oz./gal.
Temperature	160 to 180°F.
Current Density (Cathodic)	10 to 40 amp./sq.ft.
Time	15 to 60 seconds

Steel- or Copper-Base

Sodium Metasilicate	4 oz./gal.
Sodium Hydroxide	4 oz./gal.
Temperature	180 to 200°F.
Current Density (Anodic or Cathodic)	10 to 40 amp./sq.ft.
Time	15 to 60 seconds

For foaming add Naccanol NR to the extent of 0.5 per cent of the chemicals used.

Resistivity and Gravity of Freshly Prepared Plating Baths at 25°C.

	g./l.	*ohm.cm. 25°C.*	*°Bé 25°C.*
	Brass Bath		
Copper Cyanide	30	12.4	10.9
Zinc Cyanide	10		
Sodium Cyanide	56		
Sodium Carbonate	30		
	Cadmium Bath		
Cadmium Oxide	30	7.3	11.3
Sodium Cyanide	110		
	Chromium Bath		
Chromium Oxide	250	2.0	21.6
Sulfuric Acid	2.5		
	Acid Copper Bath		
Copper Sulfate	200	5.6	20.6
Sulfuric Acid	75		

	g./l.	ohm.cm. 25°C.	°Bé 25°C.
	Rochelle Copper Bath		
Copper Cyanide	26	14.3	11.0
Sodium Cyanide	35		
Rochelle Salts	45		
Sodium Carbonate	30		
	High-Efficiency Sodium Cyanide Copper Bath		
Copper Cyanide	120	6.8	22.0
Sodium Cyanide	135		
Sodium Hydroxide	30		
	High-Efficiency Potassium Cyanide Copper Bath		
Copper Cyanide	56	5.9	13.9
Potassium Cyanide	85		
Potassium Hydroxide	41		
	Acid Lead Bath		
Lead Fluoborate	105	4.5	22.6
Fluoboric Acid	40		
	Watts Nickel Bath		
Nickel Sulfate	240	19.5	21.7
Nickel Chloride	45		
Boric Acid	30		
Sulfuric Acid	0.23		
	Cyanide Silver Bath		
Silver Cyanide	40	9.1	12.7
Potassium Cyanide	60		
Potassium Carbonate	60		
	Alkaline Tin Bath		
Sodium Stannate	90	16.2	11.6
Sodium Hydroxide	8		
Sodium Acetate	15		
	Cyanide Zinc Bath		
Zinc Cyanide	60	7.7	12.0
Sodium Cyanide	23		
Sodium Hydroxide	53		
	Acid Zinc Bath		
Zinc Sulfate Crystals	240	21.0	19.5

	g./l.	*ohm.cm. 25°C.*	*°Bé 25°C.*
Ammonium Chloride	15		
Aluminum Sulfate	30		
	Alkaline Cleaner		
Sodium Carbonate	30	19.0	5.8
Trisodium Phosphate	15		
Sodium Hydroxide	8		

Single Electrode Potentials

Common plating-bath concentrations at room temperature. Hydrogen = 0.00

Electrode	*Volts Acid Baths*	*Volts Cyanide or Alkaline Baths*
Ag/Ag^{+}	—0.75	0.39
Cd/Cd^{++}	0.39	0.93
Cn/Cn^{++}	—0.36	0.65
Fe/Fe^{++}	0.20	0.25
In/In^{+++}	0.30	0.65
Ni/Ni^{++}	—0.20	0.34
Sn/Sn^{++}	0.20	
Sn/Sn^{++++}		0.79
Pb/Pb^{++}	0.15	0.55
Zn/Zn^{++}	0.79	1.15

Plating Range and Current-Density Range for Various Plating Baths

(Plating range estimated from Hull cell data)

Bath	*Recommended Operating Current Density Range amp./sq.ft.*	*Plating Range amp./sq.ft.*
Brass	3– 5	1– 60
Cadmium (Cyanide)	15– 45	1– 90
Chromium	50–450	50–500
Copper (Rochelle)	20– 60	1– 60
Copper (High-Efficiency)	10–100	1–100
Lead (Acid)	10– 80	1–130
Nickel (Bright)	15– 50	1–130
Silver	5– 15	1– 15

Bath	*Recommended Operating Current Density Range amp./sq.ft.*	*Plating Range amp./sq.ft.*
Tin (Acid)	5– 50	1– 40
Tin (Alkaline)	6– 60	1–130
Zinc (Cyanide)	5– 50	1–130

Electrochemical Formulas

$$\frac{\text{Atomic Weight}}{\text{Valence}} = \text{Equivalent Weight}$$

Equivalent weight = grams per Faraday

Faraday = 96,500 ampere-seconds = 26.81 ampere-hours

$$\frac{454}{\text{equivalent weight}} \times 26.81 = \text{ampere-hours per pound}$$

$$\frac{\text{equivalent weight}}{26.81} = \text{grams per ampere-hour}$$

density × 2.36 = grams per mil (one square foot)

$$\frac{\text{grams per mil}}{\text{grams per ampere-hour}} = \text{ampere-hours per mil (one square foot)}$$

Electrochemical Equivalents

			One square foot		One square inch
Symbol	*Valence*	*Grams per ampere-hour*	*Ampere-hours per mil*	*Grams per mil*	*Grams per mil*
Au	1	7.35	6.2	45.5	0.316
Au	3	2.45	18.6		
Ag	1	4.03	6.2	24.8	0.172
Cd	2	2.10	9.7	20.6	0.143
Cr	6	.32	51.8	16.8	0.116
Cu	1	2.37	8.9	21.0	0.146
Cu	2	1.19	17.8		
Fe	2	1.04	17.9	18.7	0.130
In	3	1.43	12.2	17.5	0.121
Ni	2	1.10	19.0	21.0	0.146
Pb	2	3.87	6.9	26.9	0.187
Rh	3	1.28	22.9	29.2	0.203
Sn	2	2.21	7.8	17.2	0.120
Sn	4	1.11	15.6		
Zn	2	1.18	14.3	16.8	0.116

Electrochemical Yields at 100% Cathode Efficiency

Metal	*Symbol*	*Atomic weight*	*Density*	*Valence*	*Grams per Faraday*	*Ampere-hours per pound*
Aluminum	Al	27.0	2.70	3	9.0	1,352
Cadmium	Cd	112.4	8.65	2	56.2	217
Chromium	Cr	52.0	7.1	6	8.7	1,398
Cobalt	Co	58.9	8.9	2	29.5	412
Copper	Cu	63.6	8.9	1	63.6	191
Copper	Cu	63.6	8.9	2	31.8	383
Gold	Au	197.2	19.3	1	197.2	62
Gold	Au	197.2	19.3	3	65.7	185
Hydrogen	H	1.008	...	1	1.0	12,170
Indium	In	114.8	7.3	3	38.3	318
Iron	Fe	55.9	7.9	2	28.0	435
Lead	Pb	207.2	11.3	2	103.6	118
Magnesium	Pb	24.3	1.74	2	12.2	997
Nickel	Ni	58.7	8.9	2	29.4	414
Oxygen	O	16.0	...	2	8.0	1,521
Platinum	Pt	195.2	21.4	4	48.8	252
Rhodium	Rh	102.9	12.4	3	34.3	355
Silver	Ag	107.9	10.5	1	107.9	113
Tin	Sn	118.7	7.3	2	59.4	205
Tin	Sn	118.7	7.3	4	29.7	410
Tungsten	W	183.9	19.1	6	30.7	396
Zinc	Zn	65.4	7.1	2	32.7	372

Glossary

The following definitions will be useful in the discussion of plating baths. These definitions are given in conjunction with the topic of this book. Some of the definitions are new and some are simple variations of a subject that requires extensive explanation such as "polarization" and terms related to polarization. The term "free cyanide," for example, is defined only for control purposes and not for the complex reactions that take place in a cyanide plating bath.

ACTIVE. Capable of reacting chemically. A metal surface on which an electrodeposit can be bonded.

ADHESION. The tendency of a plate to adhere to the basis metal. Adhesion of a plate may be satisfactory without obtaining bond or perfect adhesion.

ADDITION AGENT. Substances added to a bath to improve the quality of the deposit, extend the plating range or produce a bright plate.

AMPERE-HOUR. One ampere flowing for one hour. From the number of ampere-hours the quantity of a product formed at an electrode can be calculated.

ANODE. The positive electrode.

ANODE SLUDGE. Compounds formed on the anode due to impurities in the anode or by reaction with substances in the bath.

ATOMIC. Pertaining to the atom. Atomic or active hydrogen is formed at the cathode, which is unusually reactive before it forms molecular hydrogen.

ATOMIC WEIGHT. The weight of an element on a scale where oxygen = 16.00.

AQUEOUS. A water system.

BALANCED BATH. A plating bath in which the amount of metal supplied to the bath is approximately equal to the amount of metal removed.

BASE BOX. 218 square feet.

BASE METAL. The major metal in an alloy; such as copper in a copper-base alloy.

BASIS METAL. The metal or alloy on which an electroplate is formed.

BATH. The solution used for electroplating.

BATH VOLTAGE. The voltage from the anode to the cathode.

BAUMÉ (Bé). A specific gravity or density scale. Baumé hydrometers are used to measure the gravity of a solution.

BENT-CATHODE TEST. A plating-range test using the principle of a bent cathode to obtain a range of current densities.

BOND. A plate held to a basis metal by atomic forces. The bond strength of an electroplate is often greater than the tensile strength of the weaker of the two metals concerned.

CATHODE. The negative electrode.

CATHODE EFFICIENCY. The percentage of current used to produce the desired product at the cathode.

CHEMICAL PICKLING. Removal of the surface layer on a metal by chemical means.

COATING. A finished or protective layer.

COMPATIBLE METALS. Metals that are similar. Metals that will codeposit. Metals that will form solid solutions or compounds.

COMPLEX ION. An ion in which the metal is combined with other elements so that it is not directly charged.

CONDUCTIVITY. The ability of a bath to conduct current.

CONTAMINANTS. Impurities in a plating bath.

CONVECTION CURRENT. A flow of solution set up in a bath during electrolysis due to the change in concentration at the electrodes.

COULOMB. The unit quantity of electricity. One ampere-second.

CRYSTALLINE. Property of a surface that shows small crystal faces.

CURRENT. Electric current.

CURRENT DENSITY. The amount of current per unit area. Amperes per square foot.

CYANIDE. A chemical radical consisting of carbon and nitrogen. The free cyanide and the combined cyanide together constitute the total cyanide. However, this definition is only for analytical purposes. The combination of cyanide in the plating bath is not always known.

CYANIDE (combined). The amount of cyanide in chemical combination with a platable metal such as copper or zinc to form a complex.

CYANIDE (free). The difference between the combined and total cyanide. The amount of free cyanide as determined by chemical analysis.

Cyanide (metal). A metallic, salt of hydro cyanic acid, such as copper cyanide or zinc cyanide. Sodium cyanide is an alkali metal cyanide.

Cyanide (total). The total amount of cyanide (combined plus free) present in a plating bath. The total cyanide as determined by chemical analysis.

Decomposition Potential. The minimum potential at which a reaction at an electrode just starts.

Degreasing. Removal of grease or oil by the use of a solvent.

Deposition Potential. The potential of the cathode against the solution during electrodeposition.

Diffusion Coating. A coating formed by diffusion of one metal into another.

Drag-in. Impurities or salts introduced into a bath with the solution carried by the work into the plating line from the tanks in which it was treated prior to the bath.

Drag-out. Loss of solution from a bath, carried out by the work.

Electrochemical Series. A series of potentials of the metals based on one gram ionic weight. The approximate potentials for simple solutions. Chemical elements arranged in the order of their standard electrode potentials.

Electrodeposition. The deposition of a substance by means of electric current.

Electrode Reaction. A reaction taking place at an electrode during the passage of current.

Electroforming. Electrodeposition over a form to produce a shape that will be removed from the form.

Electrogalvanizing. Production of zinc plate by electrodeposition.

Electrolysis. Chemical decomposition by the action of electric current on an electrolyte.

Electrolyte. A substance which will conduct electricity by means of ions.

Electrolytic. Pertaining to the passage of current across an electrolyte.

Electrolytic Alkaline Cleaning. Cleaning in an alkaline bath by the passage of current. The work to be cleaned may either be cleaned anodically or cathodically.

Electrolytic Cleaning. Cleaning of a metal by electrolysis.

Electrolytic Pickling. Removal of the surface layer of a metal by electrolytic means, the metal being the anode.

ELECTRONEGATIVE. A substance which passes to the anode during electrolysis. Anodic.

ELECTROPLATING. Plating by electrodeposition.

ELECTROPOSITIVE. A substance which passes to the cathode during electrolysis. Cathodic.

ELECTROREFINING. Refining of metals by an electrolytic process.

ELECTROTINNING. Production of tin plate by electroplating.

EQUILIBRIUM POTENTIAL. The static potential between a metal and a solution of ions of the metal.

ETCH. To remove and roughen part of the surface of a metal by the action of a chemical.

EXCESSIVE POLARIZATION. Polarization in excess of that normally expected, caused by abnormal local conditions at the electrode.

FARADAY. The quantity of current required to deposit 1 gram equivalent weight; 96,500 coulombs.

FORMULATION. The composition and preparation of plating baths.

GRAM ATOMIC WEIGHT. The atomic weight of a substance in grams.

GRAM EQUIVALENT WEIGHT. The gram molecular weight divided by the valence.

GRAVITY. The density of a solution. Usually measured as Baumé gravity.

HULL TEST. A plating-range test using the principle of an inclined cathode to obtain a range of current densities.

HYDROGEN. A gas given off at the cathode during electrolysis in many electrolytic processes as a result of the electrolytic decomposition of water.

HYDROGEN OVERVOLTAGE. The voltage required for hydrogen to just begin to develop at the cathode.

INORGANIC. Chemical compounds other than organic.

ION. A charged particle formed by the loss or gain of electrons from a neutral compound in solution.

IONIZATION. Conversion of a substance into ions.

LIMITING CURRENT DENSITY. A maximum current density that cannot be exceeded without sacrificing plate quality. The current density at which a second electrode reaction begins.

METAL COMPLEX. A complex metal salt.

METAL SALT. A salt containing a metal. Metal salts may be formed by the reaction of a metal with an acid.

MIL. One thousandth of an inch (0.001 in.).

MINERAL ACIDS. Commercial inorganic acids, e.g. hydrochloric or sulfuric acid.

MOL (mole). Gram molecular weight.

NOBLE METAL. A metal that deposits easily from a plating bath in competition with a second metal. A metal that will deposit exclusively at low current density.

NOBLE-METAL SYSTEM. An alloy-plating bath system in which the analysis of the deposit is dependent on the rate of diffusion of the noble metal to the cathode.

NONAQUEOUS. Other than a water solution, such as a solution in an organic solvent or a fused salt.

NONCONDUCTOR. A substance that will not conduct current. A material used to shadow a cathode from the current during plating.

NORMAL POLARIZATION. An increase in voltage at an electrode due to the passage of current.

ORGANIC. The compounds of carbon.

OVERVOLTAGE. The minimum voltage at which a reaction at an electrode just starts to take place. A dynamic potential that is measured and can be distinguished from the decomposition potential as calculated from thermodynamic constants.

PASSIVE. Property of a surface that will not react chemically. Property of a surface on which a bonded electroplate cannot be obtained.

PICKLING. Removal of a metal surface layer with acid.

PLATE. A layer of metal over a basis metal. An electroplate.

PLATING RANGE. The entire range of current densities over which a satisfactory deposit can be obtained.

POLARIZATION. A reverse potential set up during plating that tends to resist the flow of current. Increase in voltage at an electrode due to the formation of films.

POTENTIAL SYSTEM. An alloy-plating bath system in which the analysis of the deposit is dependent on thc ratio of the two soluble metals.

PRIMARY SALT. A salt containing metal or other substances vital to the plating bath. Metal salts, e.g., cyanides are primary salts. Buffers and salts that aid anode corrosion may be considered as secondary salts.

SANDWICH EFFECT. A stressed deposit on both sides of a thin sheet which causes cracking of the sheet.

SOLVENT. A substance, usually a liquid, capable of dissolving other substances. Organic fluids used to dissolve grease or oil from the surface of metal.

SPECIFICATION PLATING. Plating to meet specifications. Often the customer, e.g., the government, requires that plating meets standards of thickness or corrosion resistance.

STANNIC. A tin compound or tin ion of valence 4. The oxidation product of stannous tin.

STANNOUS. A tin compound or tin ion of valence 2. The reduction product of stannic tin.

STRIKE. A low cathode efficiency bath used for purposes of obtaining bond of a subsequent plate.

SURFACE FILMS. Films formed on the surface of a metal by chemical action.

TRIVALENT CHROMIUM. Chromium ion formed by reduction of chromic acid during electrolysis.

VALENCE. The number of charges carried by a simple ion. The number of hydrogen atoms with which one atom of an element will combine.

VAPOR DEGREASING. Removal of grease or oil from a metal by allowing the vapors of a solvent to condense on the metal.

Index

E

F

G

L

M

N